Flight Planning

JAA ATPL Training

This book has been written and published to assist students enrolled in an approved JAA Air Transport Pilot Licence (ATPL) course in preparation for the JAA ATPL theoretical knowledge examinations. Nothing in the content of this book is to be interpreted as constituting instruction or advice relating to practical flying.

Whilst every effort has been made to ensure the accuracy of the information contained within this book, neither Jeppesen nor Atlantic Flight Training gives any warranty as to its accuracy or otherwise. Students preparing for the JAA ATPL theoretical knowledge examinations should not regard this book as a substitute for the JAA ATPL theoretical knowledge training syllabus published in the current edition of "JAR-FCL 1 Flight Crew Licensing (Aeroplanes)" (the Syllabus). The Syllabus constitutes the sole authoritative definition of the subject matter to be studied in a JAA ATPL theoretical knowledge training programme. No student should prepare for, or is entitled to enter himself/herself for, the JAA ATPL theoretical knowledge examinations without first being enrolled in a training school which has been granted approval by a JAA-authorised national aviation authority to deliver JAA ATPL training.

Contact Details:

Sales and Service Department
Jeppesen GmbH
Frankfurter Strasse 233
63263 Neu-Isenburg
Germany

Tel: ++49 (0)6102 5070
E-mail: fra-services@jeppesen.com

For further information on products and services from Jeppesen, visit our web site at:
www.jeppesen.com

ISBN 0-88487-361-7

JA310111-000

Printed in Germany

PREFACE

As the world moves toward a single standard for international pilot licensing, many nations have adopted the syllabi and regulations of the "Joint Aviation Requirements-Flight Crew Licensing" (JAR-FCL), the licensing agency of the Joint Aviation Authorities (JAA).

Though training and licensing requirements of individual national aviation authorities are similar in content and scope to the JAA curriculum, individuals who wish to train for JAA licences need access to study materials which have been specifically designed to meet the requirements of the JAA licensing system. The volumes in this series aim to cover the subject matter tested in the JAA ATPL ground examinations as set forth in the ATPL training syllabus, contained in the JAA publication, "JAR-FCL 1 (Aeroplanes)".

The JAA regulations specify that all those who wish to obtain a JAA ATPL must study with a flying training organisation (FTO) which has been granted approval by a JAA-authorised national aviation authority to deliver JAA ATPL training. While the formal responsibility to prepare you for both the skill tests and the ground examinations lies with the FTO, these Jeppesen manuals will provide a comprehensive and necessary background for your formal training.

Jeppesen is acknowledged as the world's leading supplier of flight information services, and provides a full range of print and electronic flight information services, including navigation data, computerised flight planning, aviation software products, aviation weather services, maintenance information, and pilot training systems and supplies. Jeppesen counts among its customer base all US airlines and the majority of international airlines worldwide. It also serves the large general and business aviation markets. These manuals enable you to draw on Jeppesen's vast experience as an acknowledged expert in the development and publication of pilot training materials.

We at Jeppesen wish you success in your flying and training, and we are confident that your study of these manuals will be of great value in preparing for the JAA ATPL ground examinations.

The next three pages contain a list and content description of all the volumes in the ATPL series.

ATPL Series

Meteorology (JAR Ref 050)

- The Atmosphere
- Wind
- Thermodynamics
- Clouds and Fog
- Precipitation
- Air Masses and Fronts
- Pressure System
- Climatology
- Flight Hazards
- Meteorological Information

General Navigation (JAR Ref 061)

- Basics of Navigation
- Magnetism
- Compasses
- Charts
- Dead Reckoning Navigation
- In-Flight Navigation
- Inertial Navigation Systems

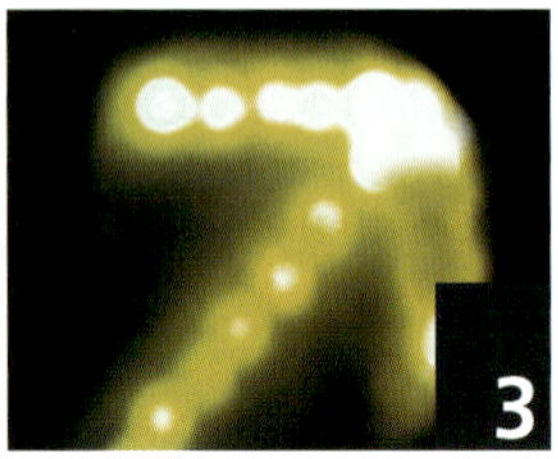

Radio Navigation (JAR Ref 062)

- Radio Aids
- Self-contained and External-Referenced Navigation Systems
- Basic Radar Principles
- Area Navigation Systems
- Basic Radio Propagation Theory

Airframes and Systems (JAR Ref 021 01)

- Fuselage
- Windows
- Wings
- Stabilising Surfaces
- Landing Gear
- Flight Controls
- Hydraulics
- Pneumatic Systems
- Air Conditioning System
- Pressurisation
- De-Ice / Anti-Ice Systems
- Fuel Systems

Powerplant (JAR Ref 021 03)

- Piston Engine
- Turbine Engine
- Engine Construction
- Engine Systems
- Auxiliary Power Unit (APU)

Electrics (JAR Ref 021 02)

- Direct Current
- Alternating Current
- Batteries
- Magnetism
- Generator / Alternator
- Semiconductors
- Circuits

Instrumentation (JAR Ref 022)

- Flight Instruments
- Automatic Flight Control Systems
- Warning and Recording Equipment
- Powerplant and System Monitoring Instruments

Principles of Flight (JAR Ref 080)

- Laws and Definitions
- Aerofoil Airflow
- Aeroplane Airflow
- Lift Coefficient
- Total Drag
- Ground Effect
- Stall
- C_{LMAX} Augmentation
- Lift Coefficient and Speed
- Boundary Layer
- High Speed Flight
- Stability
- Flying Controls
- Adverse Weather Conditions
- Propellers
- Operating Limitations
- Flight Mechanics

Performance (JAR Ref 032)

- Single-Engine Aeroplanes – Not certified under JAR/FAR 25 (Performance Class B)
- Multi-Engine Aeroplanes – Not certified under JAR/FAR 25 (Performance Class B)
- Aeroplanes certified under JAR/FAR 25 (Performance Class A)

Mass and Balance (JAR Ref 031)

- Definition and Terminology
- Limits
- Loading
- Centre of Gravity

Flight Planning (JAR Ref 033)

- Flight Plan for Cross-Country Flights
- ICAO ATC Flight Planning
- IFR (Airways) Flight Planning
- Jeppesen Airway Manual
- Meteorological Messages
- Point of Equal Time
- Point of Safe Return
- Medium Range Jet Transport Planning

Air Law (JAR Ref 010)

- International Agreements and Organisations
- Annex 8 – Airworthiness of Aircraft
- Annex 7 – Aircraft Nationality and Registration Marks
- Annex 1 – Licensing
- Rules of the Air
- Procedures for Air Navigation
- Air Traffic Services
- Aerodromes
- Facilitation
- Search and Rescue
- Security
- Aircraft Accident Investigation
- JAR-FCL
- National Law

Human Performance and Limitations (JAR Ref 040)

- Human Factors
- Aviation Physiology and Health Maintenance
- Aviation Psychology

Operational Procedures (JAR Ref 070)

- Operator
- Air Operations Certificate
- Flight Operations
- Aerodrome Operating Minima
- Low Visibility Operations
- Special Operational Procedures and Hazards
- Transoceanic and Polar Flight

Communications (JAR Ref 090)

- Definitions
- General Operation Procedures
- Relevant Weather Information
- Communication Failure
- VHF Propagation
- Allocation of Frequencies
- Distress and Urgency Procedures
- Aerodrome Control
- Approach Control
- Area Control

CHAPTER 1

Introduction to Flight Planning and Monitoring

CHAPTER 2

Introduction to CAP 697

CHAPTER 3

CAP 697 - Single Engine Piston Aeroplane (SEP 1)

CHAPTER 4

CAP 697 - Multi-Engine Piston Aircraft (MEP 1)

CHAPTER 5

CAP 697 - Medium Range Jet Transport (MRJT)

CHAPTER 6

Introduction to Jeppesen Airway Manual

CHAPTER 7

Jeppesen Airway Manual – Enroute

CHAPTER 8

Jeppesen Airway Manual – High

CHAPTER 9

Jeppesen Airway Manual - ATC, Air Reporting By Voice Communications (AIREP)

CHAPTER 10

Jeppesen Airway Manual - ATC, The Flight Plan

CHAPTER 11

Jeppesen Airway Manual – Terminal

CHAPTER 12

Jeppesen Airway Manual - Jeppesen VFR + GPS Chart, Germany ED-6

CHAPTER 13

Meteorological Messages

CHAPTER 14

Upper Air Charts

CHAPTER 15

Point of Equal Time, Point of Safe Return, and Radius of Action

CHAPTER 16

Traffic Load

CHAPTER 17

CAP 697 - Medium Range Jet Transport (MRJT) - Non-Normal Operations

Chapter 1
Introduction to Flight Planning and Monitoring

INTRODUCTION

The Flight Planning and Monitoring phase of the course is the most practical, apart from Mass and Balance. The course includes topics such as:

- CAP 697- JAR Flight Planning Manual
- Jeppesen Student Airway Manual
- Meteorological Practical
- Critical Point (Point of Equal Time) and Point of No Return

REFERENCES

The notes assume that you have both a CAP 697 and Jeppesen Airways Manual whilst you are completing each chapter. No reproductions of full diagrams are used. However, parts of charts and manuals may be used to highlight points.

NAUTICAL AIR MILES

In the CAP 697 most of the graphs are referenced to Nautical Air Miles (NAM). This is the distance flown at the TAS for a given time.

Example An aircraft is flying at a TAS of 240 knots for 45 minutes. What distance in NAM will it cover?

Using your brain, CRP 5, or a calculator, the distance covered will be **180 NAM**.

Where there is no wind component along the route that the aircraft is flying, the distance flown in NAM will be equal to the distance flown over the ground, Nautical Ground Miles (NGM).

Unfortunately life is not so easy, and the aircraft rarely encounters days when there is no wind effect.

With a headwind component, the NAM will be greater than the NGM.

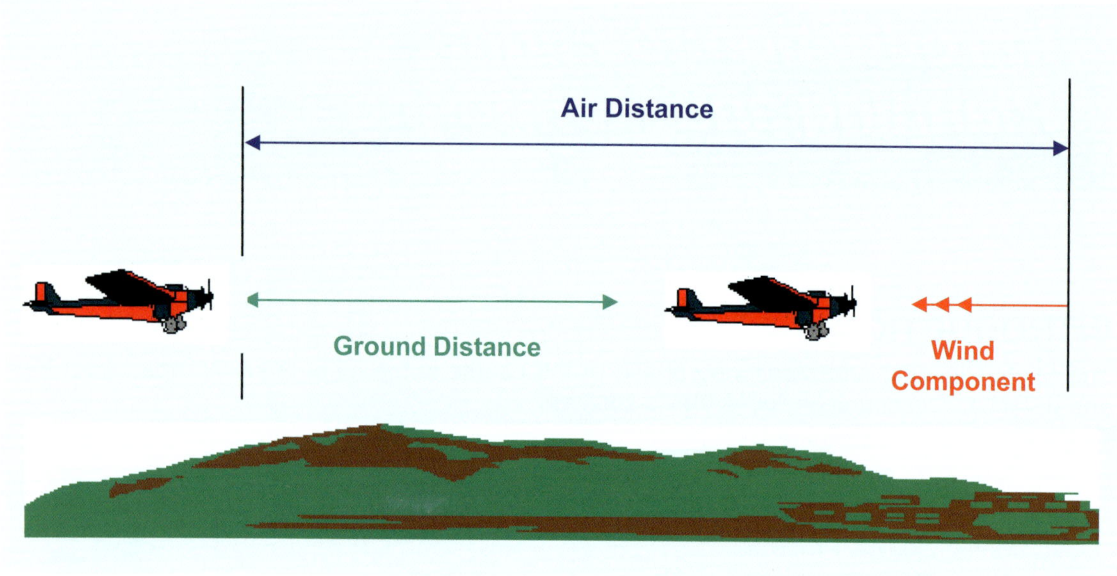

With a tailwind component, the NAM will be less than the NGM.

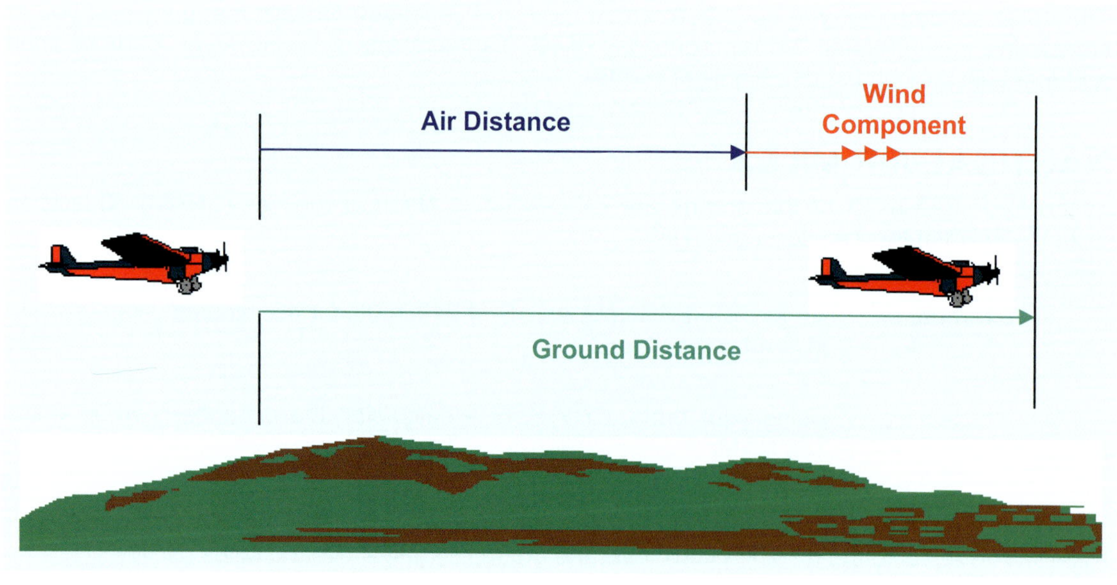

Use this simple formula to calculate the relationship:

$$\textbf{NGM = NAM x } ^{\text{GROUNDSPEED}}/_{\text{TAS}}$$

If you ever forget the formula, look on page 40 of CAP 697.

In some cases you will be given the wind component. Obtaining the groundspeed from the wind component is simple. Where a plus component is given, add the wind component to the TAS; where a minus component is given, subtract it from the TAS.

Example 1	Wind component	+ 20
	TAS	160 knots
	Groundspeed	180 knots
Example 2	Wind component	- 20
	TAS	160 knots
	Groundspeed	140 knots

Example 3 An aircraft climbs to a cruising level in 15 minutes, the distance covered is 25 NAM. The wind component is –15 kt. Calculate the NGM covered:

STEP 1	Calculate the TAS.	100 knots
STEP 2	Calculate the groundspeed.	85 knots
STEP 3	Use the formula to calculate the NGM.	

$$\text{NGM} = 25 \times {}^{85}/_{100} = 21.25 \text{ nm}$$

ANSWER THE FOLLOWING QUESTIONS:

Question	TAS	WC	Groundspeed	NGM	NAM	Time
1.		-30	150	86		
2.	210	+50			200	
3.	245		270	165	150	
4.	500	+75		300	260	
5.		-20	480	100		
6.	470	-100				25
7.		+50	350		70	
8.	375	-60			206	33
9.	200	+40		150	125	
10.		+20	420		100	15

ANSWERS TO QUESTIONS

Question	TAS	WC	Groundspeed	NGM	NAM	Time
1.	**180**	-30	150	86	**103**	**33.8**
2.	210	+50	**260**	**248**	200	**57.2**
3.	245	**+25**	270	165	150	**36.6**
4.	500	+75	**575**	300	260	**31.3**
5.	**500**	-20	480	100	**104**	**12.5**
6.	470	-100	**370**	**154**	**196**	25
7.	**300**	+50	350	**82**	70	**14.1**
8.	375	-60	**315**	**173**	206	33
9.	200	+40	**240**	150	125	**37.5**
10.	**400**	+20	420	**105**	100	15

Chapter 2
Introduction to CAP 697

INTRODUCTION

The next 4 chapters deal with the CAP 697 - Civil Aviation Authority JAR FCL Examinations Flight Planning Manual. There are no diagrams included with these chapters as it is expected that you will use the manual for all calculations. All examples include references.

FLIGHT PLANNING AND MONITORING – GENERAL NOTES

The CAP 697 that you have been given is identical to the document that candidates use in the JAR-FCL Flight Planning and Performance paper. The document follows the format of the sister documents CAPs 696 and 698 and lists three aircraft types:

Single Engine Piston (SEP 1)	Not certified under JAR 25 (Light Aeroplanes) Performance Class B
Multi Engine Piston (MEP 1)	Not certified under JAR 25 (Light Aeroplanes) Performance Class B
Medium Range Jet Transport (MRJT)	Certified under JAR 25 Performance Class A

LAYOUT

The layout of CAP 697 is the same as CAPs 696 and 698. The document is comprised of four sections:

Section I	General Notes
Section II	Single Engine Piston Aeroplane (SEP 1) – Green Paper
Section III	Multi Engine Piston Aeroplane (MEP 1) – Blue Paper
Section IV	Medium Range Jet Transport (MRJT) – White Paper

DEFINITIONS

Most of the following definitions are used in ICAO and JAA documentation. Some definitions are common use and are not used in the relevant ICAO or JAA documentation but still need to be known.

Definition	Meaning
Basic Empty Mass (Basic Mass)	The mass of an aeroplane plus standard items such as: i. Unusable fuel and other unusable fluids ii. Lubricating oil in engine and auxiliary units iii. Fire extinguishers iv. Pyrotechnics v. Emergency oxygen equipment vi. Supplementary electronic equipment
Dry Operating Mass (DOM)	The total mass of the aeroplane ready for a specific type of operation excluding all usable fuel and traffic load. The mass includes items such as: i. Crew and crew baggage ii. Catering and removable passenger service equipment iii. Potable water and lavatory chemicals iv. Food and beverages
Operating Mass (OM)	The DOM plus fuel but without traffic load
Traffic Load	The total mass of: i. Passengers ii. Baggage iii. Cargo Including any "non-revenue" load
Zero Fuel Mass	The DOM plus traffic load but excluding fuel
Maximum Zero Fuel Mass (MZFM)	The maximum permissible mass of an aeroplane with no usable fuel.
Taxi Mass	The mass of the aircraft at the start of the taxi (at departure from the loading gate).
Maximum Structural Taxi Mass	The structural limitation on the mass of the aeroplane at the commencement of taxi.
Take-Off Mass (TOM)	The mass of an aeroplane including everything and everyone contained within it at the start of the take-off run.
Performance Limited Take-Off Mass	The take-off mass subject to departure airfield limitations. It must never exceed the maximum structural limit.
Regulated TOM	The lowest of "performance limited" and "structural limited" TOM.

Definition	Meaning
Maximum Structural Take-Off Mass	The maximum permissible total aeroplane mass at the start of the take-off run.
Performance Limited Landing mass	The mass subject to the destination airfield limitations. It must never exceed the structural limit.
Maximum Structural Landing Mass	The maximum permissible total aeroplane mass on landing under normal circumstances.
Regulated Landing Mass	The lowest of "performance limited" and "structural limited" landing mass.

Note: The term "weight" is considered as having the same meaning as the term "mass".

CONVERSIONS

The following conversions are taken from the ICAO Annex, they also appear on page 3 of CAP 697.

Mass Conversion

Pounds (LB) to Kilograms (KG)	LB x 0.45359237 KG
Kilograms (KG) to Pounds (LB)	KG x 2.20462262 LB

Volumes (Liquid)

Imperial Gallons to Litres (L)	Imp Gall x 4.546092
US Gallons to Litres (L)	US Gall x 3.785412

Lengths

Feet (ft) to Metres (m)	Feet x 0.3048

Distances

Nautical Mile (NM) to metres (m)	NM x 1852

Chapter 3
CAP 697 - Single Engine Piston Aeroplane (SEP 1)

INTRODUCTION

The green pages that cover the data for the SEP 1 are found on CAP 697, pages 5 to 14. The contents are split into five areas:

1. Aeroplane description and data (CAP 697, page 6)
2. Time, fuel and distance to cruise climb (CAP 697, page 7)
3. Tables of fuel flow (CAP 697, pages 8 to 11)
4. Range profile (lean) (CAP 697, page 12)
5. Endurance profile (lean) (CAP 697, page 13)

AEROPLANE DESCRIPTION AND DATA *(CAP 697, PAGE 6)*

The SEP 1 is a monoplane with a reciprocating engine. It has a constant speed propeller with a retractable undercarriage. Assume that the undercarriage is in the correct position when making the calculations.

DETAILS

MTOM	3650 lbs
MLM	3650 lbs
Maximum fuel load	74 US gallons
Fuel Density	6 lbs per US Gallon unless advised otherwise

TIME, FUEL AND DISTANCE TO CLIMB *(CAP 697, PAGE 7)*

The graph gives the time (minutes), fuel (U.S. gallons), and distance (nautical air miles) to climb to any pressure altitude from MSL. If the departure airport is at MSL, only one entry into the graph is required. If the airfield is above MSL, make two entries and a simple calculation.

ASSOCIATED CONDITIONS

In a block to the left of the graph are the associated conditions for the climb. When "full rich" is given, this relates to the fuel/air mixture going into the engine. The terms used may be "full rich" – more fuel or "lean" – less fuel. The manifold pressure adjusts the fuel/air mixture:

The higher the manifold pressure, the more mixture being burnt.

Note that the climb speed is 110 knots which is important when actual climb distance is required.

DEPARTURE AIRFIELD AT MSL

Given the following details, calculate the time, fuel, and distance for the climb :

Airport Pressure Altitude	MSL
OAT at Cruise	+5°C
Cruise Altitude	FL 80
Climb Weight	3650 LB

STEP 1 Enter the graph at the Cruise OAT (+5°C) and move vertically to the cruise altitude (FL80).

STEP 2 Move horizontally across the graph to the Initial Climb Weight. You will see 4 climb weights to use. If another weight is given, interpolate between the figures. The one for this calculation is 3650.

STEP 3 Move vertically down to read in order:

Time	10 minutes
Fuel to Climb	3.6 US gallons
Distance to Climb	20 NAM

SEP Example 1 Given the following, calculate the time, fuel, and distance for the climb:

Airport Pressure Altitude	MSL
OAT at Cruise	+5°C
Cruise Altitude	FL 70
Climb Weight	3400 LB

SEP Example 2 Given the following, calculate the time, fuel, and distance for the climb:

Airport Pressure Altitude	MSL
OAT at Cruise	+15°C
Cruise Altitude	FL 90
Climb Weight	2600 LB

For all example questions, answers are given at the end of the chapter. Please note that your figures may not quite agree with the master answers. Some interpolation within the graph is required, so if you are within 0.5 minutes, 0.1 gallons, or 1 NAM, you need not worry.

DEPARTURE AIRFIELD AT AN ALTITUDE OTHER THAN MSL

In this calculation, allow for the notional time, fuel, and distance for the climb from MSL to the departure airfield pressure altitude. Using the example which is outlined on the graph:

OAT at Take-Off	+15°C
Airport Pressure Altitude	5653 ft
OAT at Cruise	-5°C
Cruise Altitude	11 500 ft
Climb Weight	3650 LB

STEP 1 Enter the graph at the OAT at Take-off (+15°C) and move vertically to the airport pressure altitude (5653 feet).

STEP 2 Move horizontally across the graph to the Initial Climb Weight (3650 lbs).

STEP 3 Move vertically down to read in order:

Time	6.5 minutes
Fuel to Climb	2.5 US gallons
Distance to Climb	12.5 NAM

STEP 4 Enter the graph at the OAT at Cruise (- 5°C) and move vertically to the cruise altitude (11 500 feet).

STEP 5 Move horizontally across the graph to the Initial Climb Weight (3650 lbs).

STEP 6 Move vertically down to read in order:

Time	18 minutes
Fuel to Climb	6 US gallons
Distance to Climb	36 NAM

STEP 7 Take away the figures found in STEP 3 from those in STEP 6 to find the climb:

Time	11.5 minutes (18 – 6.5)
Fuel to Climb	3.5 US gallons (6 – 2.5)
Distance to Climb	23.5 NAM (36 – 12.5)

SEP Example 3 Given the following, calculate the time, fuel, and distance for the climb:

OAT at Take-Off	+20°C
Airport Pressure Altitude	1000 ft
OAT at Cruise	+5°C
Cruise Altitude	6000 ft
Climb Weight	3650 LB

SEP Example 4 Given the following, calculate the time, fuel, and distance for the climb:

OAT at Take-Off	-10°C
Airport Pressure Altitude	4000 ft
OAT at Cruise	-20°C
Cruise Altitude	7500 ft
Climb Weight	3000 LB

ALLOWANCE FOR WIND COMPONENT

In the initial calculation, distance is appropriate to the Still Air condition. Use the formula in Chapter 1 if the distance in a wind component is required. However, before applying the formula, calculate the TAS and groundspeed in the following manner:

OAT at Take-Off	+20°C
Airport Pressure Altitude	3500 ft
OAT at Cruise	+1°C
Cruise Altitude	13 000 ft
Climb Weight	3500 LB
Wind Component	-25

Work out the time, fuel, and distance as normal:

Time	17 minutes
Fuel	5.5 gallons
Distance	36 NAM

Calculate the TAS in the following manner:

STEP 1	Take the mean pressure altitude for the climb.	**(13000 + 3500) ÷ 2 = 8250 ft**
STEP 2	Take the mean OAT for the climb.	**(1 + 20) ÷ 2 = 10.5°C**
STEP 3	Using the IAS of 110 knots taken from the climb graph, find the TAS on the CRP5. The wind component is –25, groundspeed is:	**127 knots** **102 knots**
STEP 4	Using the formula from chapter 1: NGM = NAM x $^{GS}/_{TAS}$ 36 x $^{102}/_{127}$	 **NGM = 29 nm**

The time to climb and the fuel used do not change.

SEP Example 5 Given the following, calculate the time, fuel, and distance in NAM and NGM for the climb:

OAT at Take-Off	-15°C
Airport Pressure Altitude	4500 ft
OAT at Cruise	-25°C
Cruise Altitude	9500 ft
Climb Weight	3200 LB
Wind Component	+20

SEP Example 6 Given the following, calculate the time, fuel, and distance in NAM and NGM for the climb:

OAT at Take-Off	+15°C
Airport Pressure Altitude	4000 ft
OAT at Cruise	0°C
Cruise Altitude	8500 ft
Climb Weight	3500 LB
Wind Component	-10

RECOMMENDED AND ECONOMY CRUISE POWER SETTINGS (*CAP 697, PAGES 8 TO 11)*

Four tables show the performance data for:

Table 2.2.1	25.0 in Hg (or full throttle)	2500 rpm
Table 2.2.2	25.0 in Hg (or full throttle)	2100 rpm
Table 2.2.3	23.0 in Hg (or full throttle)	2300 rpm
Table 2.3.1	21.0 in Hg (or full throttle)	2100 rpm

Data appears in the form of three tables relating to the ISA temperature deviations:

- Standard ISA Day
- ISA +20°C
- ISA -20°C

Note the conditions listed at the bottom of the page.

- The full throttle manifold pressure settings are approximate
- The shaded area on each table represents operations with full throttle

To use the table, turn to the page for the correct power setting. Use the table nearest to the temperature deviation given. If the temperature deviation falls in between, interpolation is required (e.g. (ISA -10°C)). Be sensible, only use the interpolation when the temperature deviation is up to 5° away from ±10°C.

Given the following information, calculate the fuel flow, KIAS, and KTAS:

Temperature Deviation	0°C
Altitude	FL 80
Power Setting	25" Hg @ 2500 rpm

STEP 1 Select the correct table. **Page 8 – Table 2.2.1**

STEP 2 Select the correct ISA deviation.

STEP 3 Enter at the correct pressure altitude. **FL 80**

STEP 4 Read of the values required:

Fuel Flow	79.3 pph, 13.2 gph
KIAS	152 knots
KTAS	169 knots

SEP Example 7 Given the following information, calculate the fuel flow, KIAS, and KTAS:

Temperature Deviation	-10°C
Altitude	FL 70
Power Setting	25" Hg @ 2500 rpm

SEP Example 8 Given the following information, calculate the fuel flow, KIAS, and KTAS:

Temperature Deviation	+10°C
Altitude	FL 50
Power Setting	21" Hg @ 2100 rpm

SEP Example 9 Given the following information, calculate the fuel flow, KIAS, and KTAS:

Temperature Deviation	+10°C
Altitude	FL 110
Power Setting	25" Hg @ 2100 rpm

SEP Example 10 Given the following information, calculate the fuel flow, KIAS, and KTAS:

Temperature Deviation	-20°C
Altitude	FL 120
Power Setting	23" Hg @ 2300 rpm

RANGE PROFILE *(CAP 697, PAGE 12)*

The table provides a simple and rapid means of determining the still air range for the SEP. Four power settings are illustrated. The range that is calculated from this graph includes the fuel for:

- Climb
- Cruise
- Taxi
- Run-up
- 45 minutes reserve fuel

The graph shows the range profiles for each power setting. For each power setting curve, the range initially decreases with altitude. At the level at which full throttle is reached, the range begins to increase.

Values of TAS (Kts.) are given at various levels on each range/power setting curve. Interpolate to find the TAS for a given setting. Remember that the ranges are still air distances and the wind component may affect the calculations significantly.

To calculate the range (NAM), use the following method (for ease of calculation, use the worked example):

Cruise Altitude 11 500 ft
Power Setting Full throttle, 2500 rpm

STEP 1 Enter with the altitude on the left hand side of the graph. **11 500 ft**

STEP 2 Move horizontally to the selected power setting. **Full throttle, 2500 rpm**

STEP 3 Move vertically down to read off the range in NAM. **866 NAM**

STEP 4 If the TAS is required, interpolate. Notice that the example used lies between two TAS values.

162 knots and 169 knots

By inspection, you should use a TAS of 163 knots

SEP Example 11 What is the still air range for the following conditions?

Cruise Altitude 8000 ft
Power Setting Full throttle, 2300 rpm

SEP Example 12 At what altitude can a range of 890 NAM be achieved with a power setting of Full Throttle, 2300 rpm?

SEP Example 13 What is the maximum range (NAM) that could be achieved with full throttle, 2100 rpm, and at what altitude would this occur?

ENDURANCE PROFILE *(CAP 697, PAGE 13)*

This is the amount of airborne time available for the fuel carried. The endurance graph provides a rapid method for determining the endurance of the SEP. Use the graph in a similar manner to the range profile.

Using the worked example on the graph:

Cruise Altitude 11 500 ft
Power Setting Full throttle, 2500 rpm

STEP 1 Enter with the altitude on the left hand side of the graph. **11 500 ft**

STEP 2 Move horizontally to the selected power setting. **Full throttle, 2500 rpm**

STEP 3 Move vertically down to read off the endurance. **5.39 hours**
5 hrs 23 min

STEP 4 If the TAS is required, interpolate. Notice that the example used lies between two TAS values.

162 knots and 169 knots

By inspection we should use a TAS of 163 knots

SEP Example 14 What is the endurance available with the following settings?

Cruise Altitude 10 000 ft
Power Setting Full throttle, 2300 rpm

SEP Example 15 What is the endurance and TAS for the following settings?

Cruise Altitude 11 500 ft
Power Setting Full throttle, 2300 rpm

SEP Example 16 What is the % increase in endurance when flying at an altitude of 8000 feet at 2100 rpm if power is set to 21.00 IN Hg as opposed to full throttle?

SEP EXAMPLE ANSWERS

SEP Example 1
7 minutes
2.6 US Gallons
13 NAM

SEP Example 2
7 minutes
2.6 US Gallons
13 NAM

SEP Example 3
5.5 minutes (6.5 – 1)
2 US Gallons (2.5 – 0.5)
10 NAM (12 – 2)

SEP Example 4
2.5 minutes (5.5 – 3)
0.9 US Gallons (2.1 – 1.2)
4 NAM (10 – 6)

SEP Example 5
5 minutes (8.5 – 3.5)
1.8 US Gallons (3.2 – 1.4)
10 NAM (17 – 7)
12 NGM

SEP Example 6
6 minutes (10 – 4)
2.1 US Gallons (3.6 – 1.5)
12 NAM (20 – 8)
11 NGM

SEP Example 7
Fuel flow 84.45 pph (ISA –20: 86.2, ISA: 82.7)
Fuel flow 14.1 gph (ISA –20: 14.4, ISA: 13.8)
KIAS 158 kt (ISA –20: 160.5, ISA: 155.5)
KTAS 169.5 kt (ISA –20: 169, ISA: 170)

SEP Example 8
Fuel flow 54.55 pph (ISA +20: 54.05, ISA: 55.05)
Fuel flow 9.1 gph (ISA +20: 9.0, ISA: 9.2)
KIAS 121.5 kt (ISA +20: 118.5, ISA: 124.5)
KTAS 130.5 kt (ISA +20: 129.5, ISA: 131.5)

SEP Example 9
Fuel flow 56.35 pph (ISA +20: 55.65, ISA: 57.05)
Fuel flow 9.43 gph (ISA +20: 9.3, ISA: 9.55)
KIAS 117.25 kt (ISA +20: 113.5, ISA: 121)
KTAS 137.75 kt (ISA +20: 136, ISA: 139.5)

SEP Example 10	**Fuel flow**	63.8 pph
	Fuel flow	10.6 gph
	KIAS	135 kt
	KTAS	153 kt

SEP Example 11 843 NAM

SEP Example 12 11 000 ft

SEP Example 13 905 NAM
10 800 ft

SEP Example 14 5.63 hours, which is 5 hrs 38 minutes

SEP Example 15 5.9 hours, which is 5 hrs 54 minutes
153 knots

SEP Example 16 8.2% (6.075 hours increases to 6.575 hours)

Chapter 4

CAP 697 - Multi-Engine Piston Aircraft (MEP 1)

INTRODUCTION

The blue pages that cover the data for the MEP 1 are found on CAP 697, pages 15 to 22. The contents are split into six areas:

1. Aeroplane description and data (CAP 697, page 16)
2. Time, fuel and distance to cruise climb (CAP 697, page 17)
3. Standard Temperature Range (CAP 697, pages 18)
4. Power Settings, Fuel flows and Speeds (CAP 697, page 19 and 20)
5. Endurance profile (lean) (CAP 697, page 21)
6. Fuel, Time and Distance to descend (CAP 697, page 22)

The MEP 1 and SEP 1 data sheets are very similar and are interpreted in a similar manner.

AEROPLANE DATA *(CAP 697, PAGE 16)*

The MEP 1 is a monoplane with twin reciprocating engines. The aircraft has twin counter-rotating propellers with a retractable undercarriage.

DETAILS *(CAP 697, PAGE 16)*

MTOM	4750 lb
MZFM	4470 lb
MLM	4513 lb
Maximum Fuel Load	123 US Gallons
Fuel Density	6 lb per US Gallon unless otherwise stated

POWER SETTINGS

High Speed Cruise	75%
Economy Cruise	65%
Long Range Cruise	45%

CRUISE CLIMB FUEL, TIME, AND DISTANCE TO CLIMB *(CAP 697, PAGE 17)*

Use the graph in a similar fashion to the distance, fuel, and time graph for the SEP 1. The three lines for the time, distance, and fuel to climb use a combined scale (not three different scales). Using the example on the graph:

Departure Airport Altitude	2000 ft
Departure Airport OAT	21°C
Cruise Altitude	16 500 ft
Cruise OAT	-13°C
Wind component	-20

STEP 1 Enter the graph at the departure airfield temperature and move vertically to the airfield pressure altitude. **21°C/2000 ft**

STEP 2 Move horizontally to intersect the fuel, time, and distance lines.

STEP 3 Move vertically down to read the following values:

Fuel	2 gallons
Time	3 minutes
Distance	5 NAM

STEP 4 Repeat STEPs 1 to 3 for the cruise altitude:

Fuel	15 gallons
Time	27 minutes
Distance	50 NAM

STEP 5 Subtract the values of STEP 3 from STEP 4

Fuel	13 gallons
Time	24 minutes
Distance	45 NAM

If the NGM is required use the same factoring formula as for the SEP. Calculate the mid-altitude and the mid-temperature for the climb. The IAS for the climb is listed at the top of the graph as 120 KIAS. The wind component is –20 knots.

STEP 6 Calculate the mid-altitude. **(16 500 + 2000) ÷ 2 = 9250 ft**

STEP 7 Calculate the mid-temperature. **(21 – (-13)) ÷ 2 = 17°**
21 – 17 = 4°C

STEP 8 Calculate the TAS and groundspeed. **TAS 140, groundspeed 120 kts**

STEP 9 Calculate the NGM. **NGM = 45 x $^{120}/_{140}$ = 39 NGM**

MEP Example 1 Given the following data, calculate the fuel, time, and distance for the climb:

Departure Airport Altitude	3500 ft
Departure Airport OAT	20°C
Cruise Altitude	13 000 ft
Cruise OAT	+1°C

MEP Example 2 Given the following data, calculate the fuel, time, and distance for the climb:

Departure Airport Altitude	5000 ft
Departure Airport OAT	10°C
Cruise Altitude	15 000 ft
Cruise OAT	-5°C

MEP Example 3 Given the following data, calculate the fuel, time, and distance in NAM and NGM for the climb:

Departure Airport Altitude	5000 ft
Departure Airport OAT	15°C
Cruise Altitude	15 000 ft
Cruise OAT	-15°C
Wind Component	+20

MEP Example 4 Given the following data, calculate the fuel, time, and distance in NAM and NGM for the climb:

Departure Airport Altitude	4000 ft
Departure Airport OAT	10°C
Cruise Altitude	16 000 ft
Cruise OAT	-5°C
Wind Component	-30

STANDARD TEMPERATURE RANGE *(CAP 697, PAGE 18)*

Figure 3.2 presents the range data in a graphical format. The graph is a double presentation showing:

- Two distance scales at the base:
  - Range with a 45 minute reserve at 45% power
  - Range with no reserve
- MTOM is assumed
- Standard Climb and Descent are assumed
- An allowance is made for start-up, taxi, and take-off (4.2 gallons, 25.2 lb)

Using the example in the CAP, calculate the range with and without reserve.

Cruise Altitude 16 500 ft
Power Long Range Cruise 45%

STEP 1 Enter the graph at the cruise altitude on the left hand side of the graph.

STEP 2 Move horizontally to the power selected. Then move vertically down to read the range in NAM.

Range with reserve 943 NAM
Range with no reserve 1059 NAM

MEP Example 5 Find the still air range of the aeroplane at 12 500 ft at all power settings, with and without a 45 minute reserve at 45% power.

Power Setting	Range with 45 minutes reserve at 45% power	Range with no reserve
75%		
65%		
55%		
45%		

POWER SETTING AND FUEL FLOW *(CAP 697, PAGE 19)*

Select the power setting using figure 3.3. The four percentage power columns allow selection of high speed, economy, or long range. Each percentage power column is subdivided to allow the selection of the desired rpm and manifold pressure against altitude in a standard atmosphere.

For a cruise altitude at 6000 ft and a power setting of 75%, what is the fuel flow, the rpm, and manifold air pressure?

STEP 1 Enter the table at the required % power. **Fuel flow 29 gph**

STEP 2 Read down the table to obtain the MAP.

2500 rpm 33.4" Hg
2600 rpm 32.2" Hg

There is a choice between 2500 rpm and 2600 rpm, with Manifold Pressures (MAP) given for both. Inspection of the figures at 6000 ft shows that at the LOWER rpm (preferred), MAP 34 in Hg will not be exceeded.

The table is for ISA, so make a correction according to the notes below the table.

For each 6°C above ISA	Add 1% to the MAP and the fuel flow.
For each 6°C below ISA	Subtract 1% to the MAP and the fuel flow.

MEP Example 6 Give the MAP and fuel flow for ISA conditions given:

Power	65%
RPM	2600
Altitude	6000 ft

MEP Example 7 Give the MAP and fuel flow for ISA +12°C conditions given:

Power	65%
RPM	2600
Altitude	6000 ft

SPEED POWER *(CAP 697, PAGE 20)*

This graph is used to obtain the cruise TAS for the following variables:

- Temperature
- Power setting
- Altitude

Using the example on the graph, calculate the TAS given:

Cruise OAT	-13°C
Pressure Altitude	16 500 ft
Power	55%

STEP 1	Enter the graph with the cruise OAT and go vertically to the pressure altitude.	**-13°/16 500 ft**
STEP 2	Go horizontally to the required power setting.	**55%**
STEP 3	Move vertically down to read off the TAS.	**172 knots**

MEP Example 8 Calculate the TAS given:

Cruise OAT	10°C
Pressure Altitude	11 000 ft
Power	65%

MEP Example 9 As revision, complete the following table.

ISA Deviation 0°C
Leg Distance 700 NAM
Cruise Altitude 6000 ft
RPM 2500

Power	75%	65%	55%	45%
MAP				
GPH				
TAS				
Wind Component	-20	+30	-10	+40
Groundspeed				
NGM				
Time for leg				

ENDURANCE *(CAP 697, PAGE 21)*

The next consideration is the aeroplane's endurance, given at Figure 3.5. The layout and parameters are precisely the same as for range, the only difference being that the output is Endurance in Hours.

Using the example on the graph, calculate the endurance with and without reserve:

Cruise Altitude 16 500 ft
Power 45%

STEP 1 Enter the graph at the cruise altitude on the left hand side.

STEP 2 Move horizontally to the 45% power setting lines.

STEP 3 Go vertically down to read the endurance in hours (note that the figures are in decimals of hours).

Endurance with reserve 6.16 hours
Endurance without reserve 6.91 hours

MEP Example 10 Find the endurance of an aeroplane at all power settings, with and without a 45 minute reserve at 45% power.

ISA Deviation 0°C
Cruise Altitude 12 500 ft

Power Setting	Endurance with 45 minutes reserve at 45% power	Endurance with no reserve
45%		
55%		
65%		
75%		

"Without Reserve" exceeds "With Reserve" by 45 minutes only in the 45% Power case, since in the other cases the Power level is maintained above 45% during the "Reserve Time".

FUEL, TIME, AND DISTANCE TO DESCEND *(CAP 697, PAGE 22)*

When dealing with the single engine aeroplane no descent was considered as the fuel required for the descent distance differs very little from that required for the same distance in a cruise configuration.

In twin engine aeroplanes, there is a significant difference because of the higher power and higher fuel consumption. Therefore, the table allows for the descent as a separate section of the flight. Figure 3.6 illustrates the descent data in a graphical format. The table works in exactly the same way as the climb table. Using the example, find the fuel, time, and distance for the descent:

Cruise Altitude 16 500 ft
Cruise OAT -13°C
Destination Airfield Altitude 3000 ft
Destination OAT 22°C

STEP 1 Enter the graph with the Cruise OAT and move vertically to the cruise altitude.

STEP 2 Move horizontally to the fuel, time, and distance lines.

STEP 3 Move vertically down from each line to read off the cruise values:

Fuel 6 gallons
Time 16 minutes
Distance 44 NAM

STEP 4 Repeat STEPs 1 to 3 for the destination airport.

Fuel	1 gallons
Time	3 minutes
Distance	7 NAM

STEP 5 Take the values found in STEP 4 from those found in STEP 3:

Fuel	5 gallons
Time	13 minutes
Distance	37 NAM

MEP Example 11 Find the fuel used, the time and the distance in NAM and NGM covered in a descent using the following data:

Cruise Altitude	18 000 ft
Cruise OAT	-20°C
Destination Airfield Altitude	3000 ft
Destination OAT	10°C
Wind Component	-25

MEP EXAMPLE ANSWERS

MEP Example 1

Fuel	9 gallons (12 – 3)
Time	16 minutes (22 – 6)
Distance	29 NAM (39 – 10)

MEP Example 2

Fuel	10 gallons (14 – 4)
Time	17 minutes (25 – 8)
Distance	32 NAM (46 – 14)

MEP Example 3

Fuel	9 gallons (13 – 4)
Time	16 minutes (24 – 8)
Distance	30 NAM (44 – 14) 34 NGM

MEP Example 4

Fuel	11 gallons (15 – 4)
Time	21 minutes (27 – 6)
Distance	39 NAM (50 – 11) 31 NGM

MEP Example 5

Power Setting	Range with 45 minutes reserve at 45% power	Range with no reserve
75%	**650 NAM**	**725 NAM**
65%	**768 NAM**	**865 NAM**
55%	**875 NAM**	**985 NAM**
45%	**918 NAM**	**1030 NAM**

MEP Example 6

MAP	30.3” Hg
Fuel Flow	23.3 gph

MEP Example 7

MAP	30.9” Hg
Fuel Flow	23.8 gph

MEP Example 8 180 knots

MEP Example 9

Power	75%	65%	55%	45%
MAP	33.4	31.2	26.2	21.9
GPH	29	23.3	18.7	16
TAS	171	167	152	135
Wind Component	-20	+30	-10	+40
Groundspeed	151	197	142	175
NGM	618	826	654	907
Time for leg	4 hr 06 min	4 hr 12 min	4 hr 36 min	5 hr 11 min

MEP Example 10

Power Setting	Endurance with 45 minutes reserve at 45% power	Endurance with no reserve
45%	6.32 hours	7.09 hours
55%	5.46 hours	6.09 hours
65%	4.43 hours	4.96 hours
75%	3.6 hours	4.06 hours

MEP Example 11

Fuel 4.5 gallons (6 – 1½)

Time 15 minutes (18 – 3)

Distance 41 NAM (49 – 8)

35 NGM

Chapter 5
CAP 697 - Medium Range Jet Transport (MRJT)

INTRODUCTION

The white pages that cover the data for the MRJT are found in CAP 697, pages 23 to 98. The contents are split into eight areas:

1. Aeroplane data and constants
2. Optimum altitude and short distance cruise altitude
3. Simplified flight planning
4. Holding
5. Detailed (integrated) fuel planning
6. Non-normal operations
7. Extended range operations
8. Fuel tankering

Non-normal operations, extended range operations, and fuel tankering are covered in a later chapter.

AEROPLANE DATA (*CAP 697, PAGE 24*)

The MRJT is a twin turbo-jet monoplane with a retractable undercarriage. The following structural limits apply:

Maximum Taxi (Ramp) Mass	63 060 kg
Maximum Take-Off Mass	62 800 kg
Maximum Landing Mass	54 900 kg
Maximum Zero Fuel Mass	51 300 kg
Maximum Fuel Load	5311 US Gallons 16 145 Kg (3.04 kg/US gallon)

DEFINITIONS

As a reminder:

Maximum Take-Off Mass (MTOM)	The maximum permissible total aeroplane mass at the start of the take-off run.
Maximum Zero Fuel Mass (MZFM)	The maximum permissible mass of an aeroplane with no usable fuel.
Maximum Landing Mass (MLM)	The maximum total permissible landing mass upon landing under normal circumstances.

CONSTANTS *(CAP 697, PAGE 24)*

Fuel Density	3.04 kg/US gallon 6.7 lbs/US gallon

OPTIMUM ALTITUDE (*CAP 697, PAGE 24*)

To operate a jet aeroplane at the altitude that gives the best performance, this normally means that you operate as high as possible. The performance that a pilot will be interested in may vary from flight to flight and could be any of the following:

- Best endurance
- Best range
- Best speed

A commercial aviation performance manual provides the data for a selection of cruise options. With the MRJT these options are:

- Long range cruise
- 0.74 M cruise
- 0.78 M cruise

Figure 4.2.1 is the graph for determining the Optimum Altitude for the MRJT. The graph has two curves:

- Long range cruise (LRC) or 0.74 M, and
- 0.78 M (high speed cruise)

Note that the maximum operating altitude of the MRJT is 37 000 ft.

CALCULATING THE OPTIMUM ALTITUDE (*CAP 697, PAGES 24 AND 25)*

The graph may be entered with:

- The brake release weight (this may be given as the TOM), or
- The cruise weight

Given a brake release weight of 58 250 kg, or a cruise weight of 56 800 kg, select the optimum altitude in the following manner.

STEP 1 Move vertically from the weight to the selected cruise profile.

LRC or 0.74 M	33 500 ft
0.78 M	32 700 ft

MRJT Example 1 Given the following details, calculate the optimum altitude for a 0.74 M cruise:

Brake Release Weight 62 000 kg

FUEL PENALTIES *(CAP 697, PAGE 24)*

If an aircraft is unable to operate at the optimum altitude, fuel penalties will be incurred as shown in the table below.

Off-Optimum Condition	Fuel/Mileage Penalty %	
	LRC	0.74
2000 ft above	1	1
Optimum	0	0
2000 ft below	1	2
4000 ft below	4	4
8000 ft below	10	11
12 000 ft below	15	20

The optimum altitude will increase as the fuel decreases. This can be seen in the table above. As the cruise progresses, increase the altitude to ensure that the fuel/mileage penalty is not too great.

OFF-OPTIMUM ALTITUDE

Given the following details, calculate the optimum altitude for the LRC or 0.74 M cruise. Cruise weight 58 600 kg. Using these figures, what is the fuel penalty if the aircraft is operated at 29 000 ft?

STEP 1 Calculate the optimum altitude. You are 3900 ft below optimum.

32 900 ft

STEP 2 Calculate the fuel penalty for the LRC and 0.74 M.

The fuel/mileage penalty is the same for both speeds. If the aircraft is operated at 4000 ft below optimum altitude, the penalty is 4%.

By interpolation the penalty at 3900 ft below optimum is:

LRC **3.85%**

0.74 M **3.9%**

MRJT Example 2 Given the following details below, calculate the optimum altitude and the fuel/mileage penalty for both the LRC and 0.74 M (assume that the aircraft maximum operating altitude is 36 000 ft for this question only):

Brake Release Weight 54 000 kg

Aircraft Track 145°M

SHORT DISTANCE CRUISE ALTITUDE *(CAP 697, PAGES 24 AND 25)*

For short distances, such as a positioning flight, the aeroplane may not be able to reach the optimum altitude before commencing the start of descent. For the most efficient flight, the aircraft still needs to climb as high as possible but the optimum altitude graph is inappropriate.

Figure 4.2.2 provides the information required for the calculation of Short Distance Cruise Altitudes.
Using the calculation overprinted on the graph.

Trip Distance	175 NAM
Temperature Condition	ISA +20°C
Brake Release Weight	52 000 kg

STEP 1 Enter the graph on the bottom left with the trip distance (175 NAM).

STEP 2 Move vertically to the correct temperature deviation line.

STEP 3 Move horizontally to the reference line. Follow the trade line to where the brake release weight intersects vertically.

STEP 4 Move horizontally to read the maximum pressure altitude. 28 000 ft.

MRJT Example 3 Given the details below, calculate the Short Distance Cruise Pressure Altitude:

Trip Distance	150 NAM
Temperature Condition	ISA +30°C
Brake Release Weight	55 000 kg

MRJT Example 4 Given the details below calculate, the Short Distance Cruise Pressure Altitude:

Trip Distance	200 NAM
Temperature Condition	ISA
Brake Release Weight	60 000 kg

SIMPLIFIED FUEL PLANNING *(CAP 697, PAGES 26 TO 39)*

The "Simplified Fuel Planning" charts allow a rapid determination of the:

- Estimated trip time, and
- Fuel required

From brake release to landing, the following graphs provide cruise options:

Figure 4.3.1	Long Range Cruise (Pages 28 to 30)
Figure 4.3.2	0.74 M Cruise (Pages 31 to 33)
Figure 4.3.3	0.78 M Cruise (Pages 34 to 36)
Figure 4.3.4	300 KIAS Cruise (Page 37)
Figure 4.3.5	Step Climb (Page 38)
Figure 4.3.6	Alternate Planning - LRC (Page 39)

All the graphs use the same format, and their use will be discussed later in the chapter.

ADDITIONAL ALLOWANCES *(CAP 697, PAGES 26 AND 27)*

Additional allowances are required if any of the climb, cruise, or descent schedules differ from those stated.

Cost Index Adjustment — Where a flight is planned to operate with the FMS in the "ECON" mode, adjustments are required to the LRC fuel and time. The table below accounts for the different speed profiles flown and gives both the time and fuel adjustments as a percentage:

Cost Index	Fuel Adjustment %	Time Adjustment %
0	-1	4
20	1	4
40	2	-1
60	4	-2
80	5	-3
100	7	-4
150	10	-5
200	14	-7

Ground Operations

APU Fuel Flow 115 kg/hr
Taxi Fuel 11 Kg/min

Altitude Selection For any operation away from the optimum altitude, use the table on page 24.

Cruise Trip fuel has to be increased with the AC packs at high flow. The trip fuel must also be increased for anti-icing operations:

- Engine anti-ice only: 70 kg/hr
- Engine and wing anti-ice: 180 kg/hr

Descent The simplified charts assume a descent of 0.74 M/250 KIAS with a straight-in approach. Make additional allowances for:

- For every additional minute of flaps-down manoeuvres, add 75 kg of fuel
- For engine anti-icing during the descent add 50 kg

Holding The holding fuel is determined for the table on page 40, figure 4.4

SIMPLIFIED FLIGHT PLANNING - LONG RANGE CRUISE *(CAP 697, PAGES 28 TO 30)*

The simplified long range cruise planning tables is comprised of three figures:

Figure 4.3.1A For a trip distance of 100 to 600 NGM (note that the trip distance is not given in NAM)
Figure 4.3.1B For a trip distance of 200 to 1200 NGM
Figure 4.3.1C For a trip distance of 1000 to 3000 NGM

Using figure 4.3.1A (Page 27), calculate the trip fuel and time for a LRC. This is the example on the chart):

Trip Distance	350 NGM
Cruise Altitude	29 000 ft
Estimated Landing Weight	30 000 kg
Average Wind Component	50 kts headwind
Temperature Deviation	ISA +20°C

STEP 1 Enter the graph with the trip distance.

STEP 2 Move vertically to the reference line. Correct for the 50 kt headwind by paralleling the trade lines to the 50 kt mark.

STEP 3 Move vertically to the first set of cruise altitude reference lines. From the intersection with the 29 (29 000 ft) line move horizontally to the right to the landing weight reference line.

STEP 4 Correct for the landing weight by interpolating for altitude between the two trade lines. Then move horizontally across to read the fuel.

2300 kg

STEP 5 Move vertically to the second set of cruise altitude reference lines. Move horizontally to the left to the temperature reference line.

STEP 6 Parallel the trade lines to ISA +20° and read the time.

1.1 hours (1 hr 06 min)

This method of calculation is valid for:

Figure 4.3.1	Long Range Cruise (Pages 28 to 30)
Figure 4.3.2	0.74 M Cruise (Pages 31 to 33)
Figure 4.3.3	0.78 M Cruise (Pages 34 to 36)
Figure 4.3.4	300 KIAS Cruise (Page 37)

MRJT Example 5 Using figure 4.3.1B, calculate the trip fuel and time for a LRC:

Trip Distance	1000 NGM
Cruise Altitude	25 000 ft
Estimated Landing Weight	40 000 kg
Average Wind Component	25 kts headwind
Temperature Deviation	ISA -10°C

MRJT Example 6 Using figure 4.3.2C, calculate the trip fuel and time for a 0.74 M cruise:

Trip Distance	2000 NGM
Cruise Altitude	35 000 ft
Estimated Landing Weight	30 000 kg
Average Wind Component	25 kts tailwind
Temperature Deviation	ISA +10°C

MRJT Example 7 Using figure 4.3.3A, calculate the trip fuel and time for a 0.74 M cruise:

Trip Distance	400 NGM
Cruise Altitude	25 000 ft
Estimated Landing Weight	30 000 kg
Average Wind Component	25 kts tailwind
Temperature Deviation	ISA +20°C

STEPPED CLIMB SIMPLIFIED FUEL PLANNING *(CAP 697, PAGE 38)*

The chart allows a pilot to optimise the aeroplanes performance by increasing the cruise altitude in 4000 ft steps in order to allow for the increase in optimum altitude as the aircraft burns fuel.

The graph is valid for a "step climb" of 4000 ft with the aircraft climbing to 2000 ft above the optimum altitude.

Trip fuel and time at LRC or 0.74 M is provided from brake release to touchdown. The method of use is the same as figures 4.3.1 to 4.3.4. However, brake release weight is used instead of cruise pressure altitude.

Using the example, calculate fuel and trip time:

Trip Distance	2280 NGM
Wind Component	50 kts headwind
Brake Release Weight	55 000 kg
Temperature Deviation	ISA +10°C

STEP 1 Enter the graph with the trip distance.

STEP 2 Move vertically to the reference line. Correct for the 50 kt headwind by paralleling the trade lines to the 50 kt mark.

STEP 3 Move vertically to the first set of brake release weight reference lines. From the intersection with the 55 (55 000 kg) line move horizontally to the right to read the trip fuel.

13 500 kg

STEP 4 Move vertically to the single all brake release weights reference line. Move horizontally to the left to the temperature reference line.

STEP 5 Parallel the trade lines to ISA +10° and read the time.

6.1 hours (6 hrs 06 min)

MRJT Example 8 Given the information below, use figure 4.3.5 (Page 38) to calculate fuel and trip time:

Trip Distance	2000 NGM
Wind Component	30 kts headwind
Brake Release Weight	65 000 kg
Temperature Deviation	ISA -10°C

MRJT Example 9 Given the information below, use figure 4.3.5 (Page 38) to calculate fuel and trip time:

Trip Distance	3000 NGM
Wind Component	50 kts tailwind
Brake Release Weight	50 000 kg
Temperature Deviation	ISA +10°C

ALTERNATE PLANNING *(CAP 697, PAGE 39)*

For alternate planning the table includes the:

- Missed Approach
- Climb to cruise altitude
- Cruise at LRC
- Descent and straight-in approach

For alternates further than 500 nm from the destination, use the LRC simplified flight planning charts, figures 4.3.1A to 4.3.1C.

Using the example, calculate fuel and time to the alternate:

Trip Distance	245 NGM
Wind Component	50 kts headwind
Landing Weight at Alternate	45 000 kg

STEP 1 Enter the graph with the trip distance.

STEP 2 Move vertically to the reference line. Correct for the 50 kt headwind by paralleling the trade lines to the 50 kt mark.

STEP 3 Move vertically to the landing weight at alternate reference lines. From the intersection with the 45 (45 000 kg) line, move horizontally right to read the fuel to the alternate.

1900 kg

STEP 4 Move vertically to the single all landing weights reference line. Move horizontally left to read the time to the alternate.

0.82 hours (49 minutes)

MRJT Example 10 Given the information below, use figure 4.3.6 (Page 39) to calculate the fuel and trip time to the alternate:

Trip Distance	300 NGM
Wind Component	50 kts tailwind
Landing Weight at Alternate	40 000 kg

MRJT Example 11 Given the information below, use figure 4.3.6 (Page 39) to calculate the fuel and trip time to the alternate:

Trip Distance	400 NGM
Wind Component	50 kts tailwind
Landing Weight at Alternate	60 000 kg

HOLDING FUEL PLANNING *(CAP 697, PAGE 40)*

Holding may occur for various reasons, such as:

- Weather conditions
- Congestion
- Emergency

The pilot needs to be able to calculate quickly and accurately the fuel within the hold.

For the MRJT, the holding fuel table figure 4.4 can be found on Page 40. The chart is based on two assumptions:

- The aircraft will hold in a racetrack pattern
- The aircraft will fly at minimum drag speed – 210 knots

If a straight and level hold is used, the table values can be reduced by 5%.

When interpolation is required from the table, note that the figures are fuel flow in Kg/hr.

Given the conditions below, what is the required holding fuel:

Aircraft Weight	53 000 kg
Holding Altitude	8000 ft
Holding Time	30 minutes

STEP 1 Move to the weight columns for 54 000 kg and 52 000 kg.

STEP 2 Select the two altitudes nearest 8000 ft (5000 and 10 000 ft).

STEP 3 Calculate the 53 000 kg fuel flow figures for 5000 ft and 10 000 ft.

5000 ft	**2420 kg/hr**
10 000 ft	**2380 kg/hr**

STEP 4 Calculate the 8000 ft fuel flow by interpolation and then the fuel required.

8000 ft	**2396 kg/hr**
Holding Fuel	**1198 kg**

MRJT Example 12 Given the conditions below, what is the required holding fuel:

Aircraft Weight	43 000 kg
Holding Altitude	18 000 ft
Holding Time	40 minutes

MRJT Example 13 Given the conditions below, what is the required holding fuel:

Aircraft Weight	51 000 kg
Holding Altitude	23 000 ft
Holding Time	20 minutes

DETAILED FUEL PLANNING *(CAP 697, PAGES 40 TO 98)*

The detailed fuel planning information available includes:

Figure 4.5.1	Enroute Climb	4 tables (pages 41 to 44)
Figure 4.5.2	Wind Range Correction Graph	1 table (page 45)
Figure 4.5.3.1	Long Range Cruise	11 tables (pages 47 to 57)
Figure 4.5.3.2	0.74 M Cruise	17 tables (pages 58 to 74)
Figure 4.5.3.3	0.78 M Cruise	6 tables (pages 75 to 80)
Figure 4.5.3.4	Low Level Cruise – 300 KIAS	8 tables (pages 81 to 88)
Figure 4.5.4	Descent Tables	2 tables (page 89)
Figure 4.6.1	Non-Normal Operation – Gear Down Ferry Flight	1 table (page 90)
Figure 4.7.1a	Critical Fuel Reserve – One Engine Inoperative	1 table (page 92)
Figure 4.7.1b	Critical Fuel Reserve – All Engines Operative	1 table (page 93)
Figure 4.7.2	Area of Operation –Diversion Distance One Engine Inoperative	1 table (page 94)
Figure 4.7.3	In-Flight Diversion (LRC) One Engine Inoperative	1 table (page 95)
Figure 4.8.1	Fuel Tankering (LRC and 0.74 M)	1 table (page 97)
Figure 4.8.2	Fuel Price Differential	1 table (page 98)

ENROUTE CLIMB *(CAP 697, PAGES 40 TO 44)*

Four climb tables are given for different ISA temperature deviations:

- ISA –6°C to –15°C
- ISA –5°C to +5°C
- ISA +6°C to +15°C
- ISA +16°C to +25°C

The fuel and time in the tables are from brake release, and the distance from 1500 ft with a climb airspeed of 280 kts/0.74 M.

Any stated TAS is the average for the climb and is used to correct the still air distance to NGM. Find the formula on page 40.

Given the data below, calculate the enroute climb data:

Brake Release Weight	62 000 kg
Airport Elevation	MSL
Cruise Level	33 000 ft
ISA Deviation	-10°

STEP 1 Select the correct table by checking the ISA deviation. The ISA deviation is within the range –6°C to –15°C.

Figure 4.5.1 (page 41)

STEP 2 Select the cruise altitude from the left hand column – 33 000 ft.

STEP 3 Move right to the brake release weight column – 62 000 kg.

STEP 4 Read the data for the climb.

Time	**19 minutes**
Fuel	**1550 kg**
Distance	**104 nm**
TAS	**374 knots**

MRJT Example 14 Given the information below, calculate the enroute climb data:

Brake Release Weight	66 000 kg
Airport Elevation	MSL
Cruise Level	29 000 ft
ISA Deviation	+10°

Where the elevation of the airport is above mean sea level, there is a fuel adjustment. You will find this at the bottom of all of the tables. Make sure to use the adjustment relevant to that table as the fuels do differ.

MRJT Example 15 Given the information below, calculate the enroute climb data:

Brake Release Weight	59 000 kg
Airport Elevation	3000 ft
Cruise Level	35 000 ft
ISA Deviation	-13°C
Wind Component	-30 knots

Remember to use the fuel adjustment at the bottom of the table and use the wind component to calculate the NGM.

MRJT Example 16 Given the information below, calculate the enroute climb data:

Brake Release Weight	63 000 kg
Airport Elevation	5000 ft
Cruise Level	29 000 ft
OAT	-55°C
Wind Component	+30 knots

WIND RANGE CORRECTION *(CAP 697, PAGE 45)*

This graph is used for the conversion of NGM to NAM. The graph is used in conjunction with the detailed fuel planning tables found on pages 47 to 88. Note that the table uses average TAS.

Given the details below, calculate the air distance in NAM:

Average TAS	450 knots
Leg Distance	4000 NGM
Wind Component	50 kts tailwind

STEP 1 Enter the graph in the bottom left hand corner with the average TAS.

STEP 2 Move vertically upwards to the 50 kt tailwind line. The tailwind and headwind lines are clearly marked to the right.

STEP 3 Move horizontally right to the 400 reference line.

STEP 4 Move vertically down to read the NAM.

NAM 3600

By using the formula, the NAM is calculated as 3600 NAM, which is the same value as calculated by using the graph. This is not always the case as minor errors can be found between the values. The JAA will specify if the wind range correction graph is to be used.

MRJT Example 17 Given the details below, calculate the air distance in NAM.

Average TAS	400 knots
Leg Distance	350 NGM
Wind Component	50 kts headwind

MRJT Example 18 Given the details below, calculate the air distance in NAM.

Average TAS	350 knots
Leg Distance	2500 NGM
Wind Component	150 kts headwind

INTEGRATED RANGE *(CAP 697, PAGES 46 TO 88)*

Use this section to plan the cruise. The tables are identical in use. The principle used is based on "differences" between two gross weights representing the weight of fuel used. The corresponding difference in tabulated distance represents the still air distance available for that weight of fuel used.

Using figure 4.5.3.1 (Page 47), the table represents the LRC at 27 000 ft. The left hand column represents the gross weight in thousands of kilograms. For convenience, across the top of the table weights are tabulated in hundreds of kilograms. This removes the need for some interpolation.

The second column from the left gives the TAS for the weight. Note that this reduces as the aircraft weight reduces. The remainder of the columns represent the NAM the aircraft could fly at that weight.

For example, in the bottom right hand corner of the table the figure for 67 900 kg is 5687 NAM. This is the still air distance the aircraft could fly with zero fuel at 35 000 kg.

Given the details below, calculate the fuel used:

Leg Distance (NAM) 3000 NAM
Gross Weight 62 700 kg

STEP 1 Enter the table in the left-hand column at 62 000 kg. Move right to the 700 kg column to make 62 700 kg. Read of the initial distance in NAM.

4910 NAM

STEP 2 The leg distance is 3000 NAM, subtract this from the start NAM of 4910 NAM.

4910 – 3000 = 1910 NAM

This figure is the NAM the aircraft could fly to zero fuel.

STEP 3 Enter the Cruise Distance Nautical Air Miles Columns and find the nearest figure to 1910. This occurs at:

Gross Weight 44 800 kg 1903 NAM
Gross Weight 44 900 kg 1922 NAM

For more accuracy 1910 is approximately half way between the two figures so the end gross weight will be:

Gross Weight 44 850 kg

STEP 4 Subtract this gross weight from the start gross weight to give the fuel used for the leg.

62 700 – 44 850 = 17 850 kg

TEMPERATURE DEVIATION

At the bottom of each table there are four required adjustments for deviations from ISA. These are not the same for each table, using figure 4.5.3.2 (page 70), 0.74 M Cruise at 33 000 ft.

Increase Fuel Required By	0.6% per 10°C above ISA
Decrease Fuel Required By	0.6% per 10°C below ISA
Increase TAS by	1 knot per degree C above ISA
Decrease TAS by	1 knot per degree C below ISA

Given the details below, calculate the NAM for the two legs and the fuel used for each leg:

Cruise Speed	0.74 M
Cruise Altitude	33 000 ft
Gross Weight	53 500 kg
ISA Deviation	0
Route	

	NGM	**Wind Component**
A – B	240	-20
B – C	370	-30

STEP 1 Select the correct cruise table for 0.74 M.

Figure 4.5.3.2 **Page 70**

STEP 2 Using the TAS from the top right of the table (430 kts) calculate the NAM for each leg.

A – B **252 NAM**
B – C **398 NAM**

STEP 3 Using the gross weight at the start of leg A – 53 500 kg find the start value for the cruise distance:

Start NAM **3796 NAM**

STEP 4 Subtract the first leg NAM.

3796 – 252 = 3544 NAM

This equates to an end of leg gross weight of **52 100 kg** (go to the nearest 100 kg).

The leg fuel is therefore:

53 500 – 52 100 = 1400 kg

STEP 5 Using the NAM at the end of the first leg subtract the second leg distance:

3544 – 398 = 3146

This equates to an end of leg gross weight of **50 000 kg** (go to the nearest 100 kg).

The leg fuel is therefore:

52 100 – 50 000 = 2100 kg

If there is any deviation from ISA, apply the adjustments from the bottom of the table.

MRJT Example 19 Given the details below, calculate the leg fuels:

Cruise Speed 0.74 M
Cruise Altitude 29 000 ft
Gross Weight 61 500 kg
ISA Deviation +10°C
Route

	NGM	Wind Component
A – B	750	+20
B – C	450	-30

MRJT Example 20 Given the details below, calculate the leg fuels:

Cruise Speed 0.78 M
Cruise Altitude 31 000 ft
Gross Weight 63 700 kg
ISA Deviation -25°C
Route

	NGM	Wind Component
A – B	650	-50
B – C	850	+40

If you use the Long Range Cruise tables, use the TAS from the second column. Remember that if you use this table in conjunction with the Wind Correction Graph on Page 45, you must calculate an average TAS.

DESCENT *(CAP 607, PAGE 89)*

Two tables are provided for:

- 0.74 M/250 KIAS
- 0.70 M/280/250 KIAS

Both descents are based upon idle thrust with allowances included for a straight-in approach with gear down. Remember that the descent fuel must be adjusted in the following cases:

- If engine anti-icing is used in the descent, add 50 kg.
- For every additional minute of flaps-down manoeuvre, add 75 kg.

Given the details below calculate the time, fuel, and distance NAM for the descent:

Landing Weight	47 500 kg
Descent Speed	0.74 M/250 KIAS
Cruise Altitude	30 000 ft

STEP 1 Enter the table in the left hand column, the pressure altitude. Read off the time, fuel used, and distance NAM:

Time	**19.5 minutes**
Fuel Used	**277.5 kg (275 or 280 kg is acceptable)**
Distance	**92 NAM**

Where the NGM are required, the mid-altitude temperature is required to work out the TAS:

Using the example above, assume that the temperature deviation is 0°C with a wind component of +50 knots. The temperature at 15 000 ft (half-altitude) is –15°C. This results in a TAS of 315 knots using a descent KIAS of 250 knots. Use this TAS to calculate the NGM. In this case 107 NGM.

MRJT Example 21 Given the details below calculate the time, fuel, and distance NAM for the descent:

Landing Weight	37 500 kg
Descent Speed	0.74 M/250 KIAS
Cruise Altitude	26 000 ft
Temperature Deviation	-5°C
Wind Component	-20 knots

MRJT EXAMPLE ANSWERS

MRJT Example 1	32 200 ft
MRJT Example 2	**Optimum Altitude** 35 100 ft Aircraft Track 145°M gives a semi-circular altitude of either FL 330 or FL 370. With the assumption that the maximum operating altitude is 36 000 ft, the correct semi-circular is FL 330. This is 2100 ft below optimum altitude, by interpolation: **LRC fuel/mileage penalty 1.15%** **0.74 M cruise 2.1%**
MRJT Example 3	24 500 ft
MRJT Example 4	29 500 ft
MRJT Example 5	**Fuel** 6600 kg **Time** 2.8 hrs
MRJT Example 6	**Fuel** 8700 kg **Time** 4.5 hrs
MRJT Example 7	**Fuel** 2680 kg **Time** 0.9 hrs
MRJT Example 8	**Fuel** 13 200 kg **Time** 5.3 hrs
MRJT Example 9	**Fuel** 12 600 kg **Time** 6.3 hrs
MRJT Example 10	**Fuel** 1750 kg **Time** 0.79 hrs
MRJT Example 11	**Fuel** 2800 kg **Time** 0.99 hrs
MRJT Example 12	**Fuel** 1280 kg
MRJT Example 13	**Fuel** 736 kg
MRJT Example 14	**Time** 18 minutes **Fuel** 1600 kg **Distance** 98 NAM **TAS** 376 knots

MRJT Example 15	**Time**	19.5 minutes
	Fuel	1475 kg
	Distance	110 NAM
		101 NGM
	TAS	379 knots

MRJT Example 16	**Time**	15.5 minutes
	Fuel	1225 kg
	Distance	79 NAM
		86 NGM
	TAS	359.5 knots

MRJT Example 17 400 NAM (calculated 400 NAM)

MRJT Example 18 4400 NAM (calculated 4375 NAM)

MRJT Example 19 Figure 4.5.3.2 (page 66)
Basic TAS from table 438 knots
Temperature corrected TAS 448 knots

Leg	NAM	Start NAM	End NAM	Start Gross Weight	End Gross Weight	Fuel Used (no temp correction)	Fuel Used
A - B	717	4791	4074	61 500 kg	57 100 kg	4400 kg	4426 kg
B - C	482	4074	3592	57 100 kg	54 300 kg	2800 kg	2816 kg

MRJT Example 20 Figure 4.5.3.3 (page 76)
Basic TAS from table 460 knots
Temperature corrected TAS 435 knots

Leg	NAM	Start NAM	End NAM	Start Gross Weight	End Gross Weight	Fuel Used (no temp correction)	Fuel Used
A - B	734	4939	4205	63 700 kg	58 800 kg	4900	4974 kg
B - C	778	4205	3427	58 800 kg	53 900 kg	4900	4974 kg

MRJT Example 21	**Time**	18.5 minutes
	Fuel	265 kg
	Distance	72 NAM
		67 NGM

Chapter 6
Introduction to Jeppesen Airway Manual

INTRODUCTION

The *Jeppesen Student Pilot Route Airway Manual* is used for the Flight Planning exam. You will need to know how and where to extract information quickly and accurately. This chapter has been produced to help you navigate through the structure of the manual.

The manual you have is the one that you will use in the examinations.

DO NOT MARK IT OR HIGHLIGHT MATERIAL IN ANY WAY

DO NOT CHANGE THE ORDER OF ANY ITEM

Note: The charts are not current and are **NOT** to be used for navigation purposes. The manual is a training aid for the JAR ATPL examinations only.

The manual is split into five sections:

- Introduction
- Enroute
- High
- Air Traffic Control
- Terminal

INTRODUCTION TO THE JEPPESEN MANUAL

TABLE OF CONTENTS

This document uses those parts of the Jeppesen Manual that are relevant to the Flight Planning examination. There are no pages missing. The original page numbers are retained and this leaves what appear to be gaps in the manual. The Chart Glossary, for instance, runs from Pages 1 to 14, whereas the next page in the manual is Abbreviations on Page 41.

CHART GLOSSARY *(JEPPESEN MANUAL, PAGES 1 TO 14)*

The chart glossary provides definitions commonly used in aviation publications. You will find any required definition in this section. No explanation of the definitions is given in this document. For examination purposes it will be beneficial for you to read and understand the definitions listed.

ABBREVIATIONS *(JEPPESEN MANUAL, PAGES 41 TO 45)*

Any abbreviation that is used in the Jeppesen Manual is found in this section.

ENROUTE CHART LEGEND – GENERAL *(JEPPESEN MANUAL, PAGES 51 TO 70)*

Jeppesen Enroute charts are compiled and constructed using:

- A Lambert Conformal Conic Projection, and
- All available aeronautical and topographical references

Each chart uses:

- Code letters for the world areas covered
- Letters giving the altitude coverage
- A number relevant to the individual chart

e.g. P(H/L)2

This is a chart for the Pacific region that covers both high and low level altitude operations and is the second chart in the series.

Go to the enroute section of the manual and choose the first chart – E(LO) **1A**

CHART CODE

This is Chart 1A in the European series and is for low altitude operations.

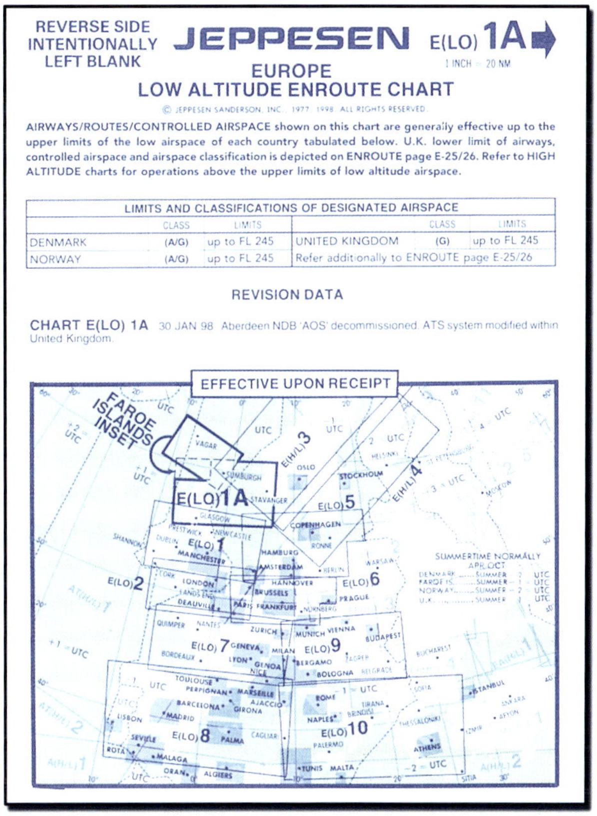

AREA OF COVERAGE

The box on the front of the chart shows a map of Europe and the code numbers for chart coverage. If the coverage of Spain and Portugal were required chart, E(LO)8 would be used.

A box highlights the area of coverage of the chart. Note that in the area coverage for the chart there are five names:

- *Glasgow*
- *Prestwick*
- *Sumburgh*
- *Stavanger*
- *Vagar*

At the top of the panel is a box with the words **EFFECTIVE UPON RECEIPT**. Chart effective dates are given when significant changes become effective on the date shown.

ADDITIONAL INFORMATION

Included on this first page are details of the coverage of low level airspace for the countries for which the chart is effective. This includes the classes of airspace involved.

Listed above the area coverage box is the revision data since the last published chart. In this case, the Aberdeen NDB was decommissioned effective 30 January 1998. This revision data is supplemented by enroute chart NOTAMS when significant changes occur between revision dates. Chart revision dates are always on a Friday.

The scale of the chart is underneath the chart code.

COMMUNICATIONS

Communications are shown in two ways:

- On the chart, or
- Tabulated on the end folds under communications

Terminal communications are provided on the LO charts.

The communications required for each airfield on the chart are listed on this panel.

e.g. Perth

PERTH, U.K. EGPT p1D
Scone. **Perth *Rdo** 119.8

- The top line gives the airfield name, country, and ICAO identifier. To the right of the identifier is a chart code p1D (an explanation is given later in this chapter).
- The lower line gives the identifying name in regular text and the radio callsign in bold text with the frequency. The asterisk means that the airfield is part-time operation only. An explanation of the information is given in the text above the listed airfields.

TRANSPONDER SETTINGS

Below the airfield list are the required SSR procedures and the cruising levels used in the countries of coverage. Note that the Transponder Settings refers to pages ENROUTE E-17/18, which are not included in this manual.

CRUISING LEVELS

Beneath the Transponder settings are the relevant cruising levels. Notice on this chart the quadrantal procedures for the United Kingdom and the semi-circular system for the other countries on the chart.

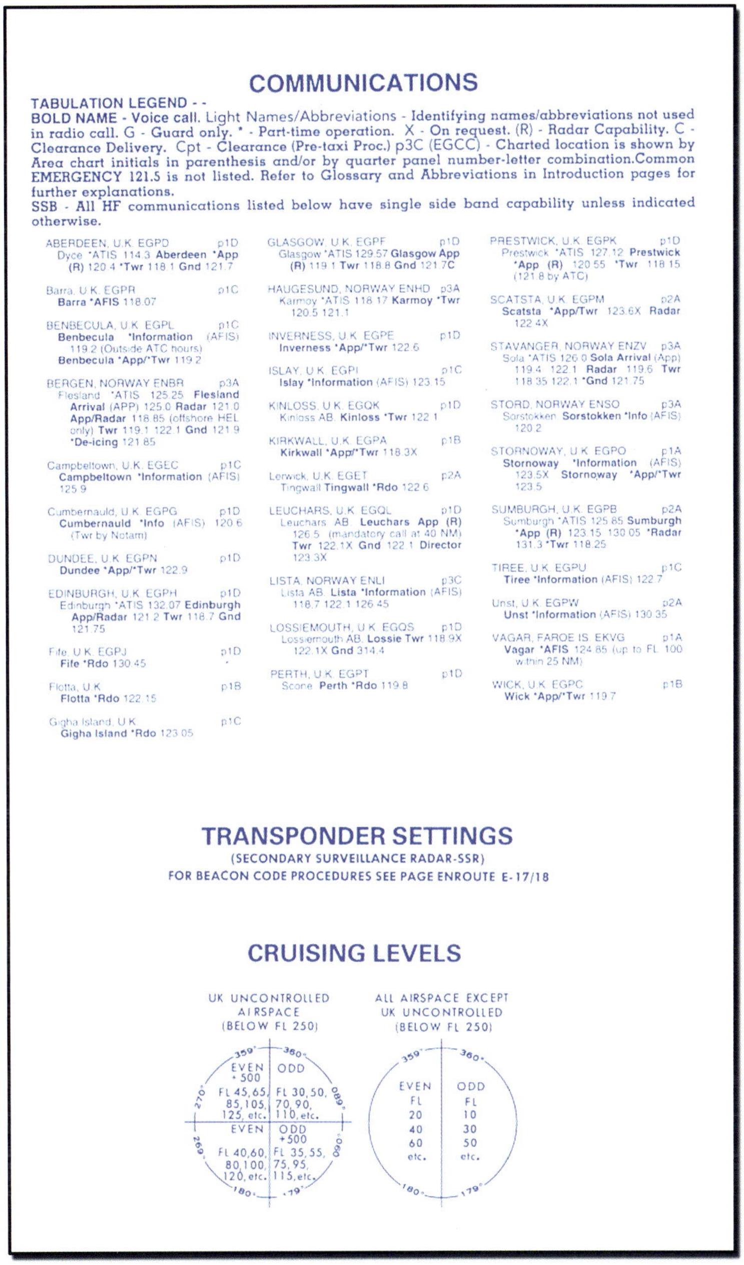

COMMUNICATIONS

TABULATION LEGEND - -
BOLD NAME - Voice call. Light Names/Abbreviations - Identifying names/abbreviations not used in radio call. G - Guard only. * - Part-time operation. X - On request. (R) - Radar Capability. C - Clearance Delivery. Cpt - Clearance (Pre-taxi Proc.) p3C (EGCC) - Charted location is shown by Area chart initials in parenthesis and/or by quarter panel number-letter combination.Common EMERGENCY 121.5 is not listed. Refer to Glossary and Abbreviations in Introduction pages for further explanations.
SSB - All HF communications listed below have single side band capability unless indicated otherwise.

ABERDEEN, U.K. EGPD p1D
Dyce *ATIS 114.3 **Aberdeen *App (R)** 120.4 ***Twr** 118.1 **Gnd** 121.7

Barra, U.K. EGPR p1C
Barra *AFIS 118.07

BENBECULA, U.K. EGPL p1C
Benbecula *Information (AFIS) 119.2 (Outside ATC hours)
Benbecula *App/*Twr 119.2

BERGEN, NORWAY ENBR p3A
Flesland *ATIS 125.25 **Flesland Arrival** (APP) 125.0 **Radar** 121.0 **App/Radar** 118.85 (offshore HEL only) **Twr** 119.1 122.1 **Gnd** 121.9 ***De-icing** 121.85

Campbeltown, U.K. EGEC p1C
Campbeltown *Information (AFIS) 125.9

Cumbernauld, U.K. EGPG p1D
Cumbernauld *Info (AFIS) 120.6 (Twr by Notam)

DUNDEE, U.K. EGPN p1D
Dundee *App/*Twr 122.9

EDINBURGH, U.K. EGPH p1D
Edinburgh *ATIS 132.07 **Edinburgh App/Radar** 121.2 **Twr** 118.7 **Gnd** 121.75

Fife, U.K. EGPJ p1D
Fife *Rdo 130.45

Flotta, U.K. p1B
Flotta *Rdo 122.15

Gigha Island, U.K. p1C
Gigha Island *Rdo 123.05

GLASGOW, U.K. EGPF p1D
Glasgow *ATIS 129.57 **Glasgow App (R)** 119.1 **Twr** 118.8 **Gnd** 121.7**C**

HAUGESUND, NORWAY ENHD p3A
Karmoy *ATIS 118.17 **Karmoy *Twr** 120.5 121.1

INVERNESS, U.K. EGPE p1D
Inverness *App/*Twr 122.6

ISLAY, U.K. EGPI p1C
Islay *Information (AFIS) 123.15

KINLOSS, U.K. EGQK p1D
Kinloss AB. **Kinloss *Twr** 122.1

KIRKWALL, U.K. EGPA p1B
Kirkwall *App/*Twr 118.3X

Lerwick, U.K. EGET p2A
Tingwall **Tingwall *Rdo** 122.6

LEUCHARS, U.K. EGQL p1D
Leuchars AB. **Leuchars App (R)** 126.5 (mandatory call at 40 NM) **Twr** 122.1X **Gnd** 122.1 **Director** 123.3X

LISTA, NORWAY ENLI p3C
Lista AB. **Lista *Information** (AFIS) 118.7 122.1 126.45

LOSSIEMOUTH, U.K. EGQS p1D
Lossiemouth AB. **Lossie Twr** 118.9X 122.1X **Gnd** 314.4

PERTH, U.K. EGPT p1D
Scone **Perth *Rdo** 119.8

PRESTWICK, U.K. EGPK p1D
Prestwick *ATIS 127.12 **Prestwick *App (R)** 120.55 ***Twr** 118.15 (121.8 by ATC)

SCATSTA, U.K. EGPM p2A
Scatsta *App/Twr 123.6X **Radar** 122.4X

STAVANGER, NORWAY ENZV p3A
Sola *ATIS 126.0 **Sola Arrival** (App) 119.4 122.1 **Radar** 119.6 **Twr** 118.35 122.1 ***Gnd** 121.75

STORD, NORWAY ENSO p3A
Sorstokken **Sorstokken *Info** (AFIS) 120.2

STORNOWAY, U.K. EGPO p1A
Stornoway *Information (AFIS) 123.5X **Stornoway *App/*Twr** 123.5

SUMBURGH, U.K. EGPB p2A
Sumburgh *ATIS 125.85 **Sumburgh *App (R)** 123.15 130.05 ***Radar** 131.3 ***Twr** 118.25

TIREE, U.K. EGPU p1C
Tiree *Information (AFIS) 122.7

Unst, U.K. EGPW p2A
Unst *Information (AFIS) 130.35

VAGAR, FAROE IS. EKVG p1A
Vagar *AFIS 124.85 (up to FL 100 within 25 NM)

WICK, U.K. EGPC p1B
Wick *App/*Twr 119.7

TRANSPONDER SETTINGS

(SECONDARY SURVEILLANCE RADAR-SSR)
FOR BEACON CODE PROCEDURES SEE PAGE ENROUTE E-17/18

CRUISING LEVELS

THE CHART

Open the chart. Shown below is an inset from the top of the chart.

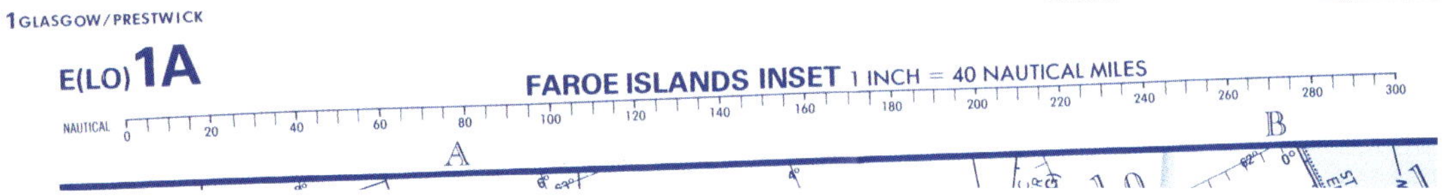

At the top of the chart in the left hand corner is:

1 GLASGOW/PRESTWICK

This refers to the place names found on the front of the chart and also the panel number. In this case panel 1.

If the chart is fully opened, three place names can be found along the top of the chart:

1. Glasgow/Prestwick
2. Sumburgh
3. Stavanger

The only name missing from the front coverage chart is Vagar. This is found under Glasgow/Prestwick as an inset for the Faroe Islands.

These numbers refer to panel numbers. Notice, underneath the scale at the top of the chart, the letters A and B. These letters are repeated in the order ABABA.

At the bottom of the chart are the letters CDCDC.

These letters, when used with a panel number, give a quick reference system for finding the listed airfields referred to on the communications panel.

Open the chart so that only Scotland can be seen. This has opened 4 panels (Each fold is a panel). On the top of the chart, you should only have two letters AB and on the bottom of the chart the two letters CD.

Using the communications listing for Perth the identifying chart code is p1D. To use this code:

1. p1 refers to the panel.
   At the top left corner of the chart is **1 GLASGOW/PRESTWICK**.
   This means that the panels opened are panel 1.
2. D refers to the bottom right panel.
3. Perth airfield will be found in this panel

SCALE

The scale of the chart is listed on the information panel under the chart code. The general scale for this chart is 1 inch = 20 nm. Any deviations are listed on the chart. For instance, the Faroe Islands inset is to a scale of 1 inch = 40 nm. For convenience, you can find the scale at the top and sides of the chart.

MEASUREMENTS *(unless otherwise indicated)*

Measurement	
Bearings and Radials	magnetic
Enroute distances	nautical miles
Vertical measurements of elevation	feet above mean sea level
Enroute altitudes	feet above mean sea level (based on QNH altimeter setting), or expressed as Flight Levels (FL) based on the standard altimeter setting of 29.92 inches of Mercury or 1013.2 hPa
Time	UTC unless labeled LT

CONGESTION

For large metropolitan areas, complete off-airway information is not always shown on the enroute chart. These areas have Area Charts of a larger chart scale with all the information required. The area chart should be used by flights arriving or departing any airport within an area chart. (Area charts will be discussed later with arrivals and departures). The area charts available are identified by

- shaded blue areas on the cover panel area of coverage chart, or
- a heavy dashed line with location name and airport identifier on the enroute chart.

CHART SYMBOLS *(JEPPESEN MANUAL, PAGES 52 TO 74)*

The chart symbols used

- on two colour charts are either blue or green.
- on single colour charts all are blue.

These pages should be used with reference to the chart as a full understanding of the symbology is required.

All compass roses are aligned to magnetic north. Values of variation can be found along the edges of the chart. Note that NDBs do not have a compass rose; they have a variation arrow.

CLASS B AIRSPACE CHART LEGEND *(JEPPESEN MANUAL, PAGE 75)*

Slight differences can be found when using the United States Low Altitude charts. An example is shown on this page. The chart depicts the horizontal and vertical limits of Class B airspace in the USA. The Class B Airspace Chart includes only the general IFR and VFR procedures appropriate to the area.

SID AND STAR LEGEND *(JEPPESEN MANUAL, PAGES 81 TO 82)*

Further explanation of these pages will be given when the SIDs and STARs are discussed in a later chapter. Like the chart legends, the legends on these pages need to be learnt.

SID AND STAR AND PROFILE DESCENT LEGEND *(JEPPESEN MANUAL, PAGES 83 TO 84)*

Further explanation of these pages will be given when the SIDs and STARs are discussed in a later chapter. Like the chart legends, the legends on these pages need to be learnt. Note that Page 84 refers to USA FAA only. New York is included in the Terminal section of this manual.

APPROACH CHART LEGEND *(JEPPESEN MANUAL, PAGES 101 TO 148)*

In addition to the SIDs and STARs, the approach charts will also be explained in a later chapter. The legends for these charts need to be understood as above.

ICAO RECOMMENDED AIRPORT SIGNS AND RUNWAY MARKINGS *(JEPPESEN MANUAL, PAGES 161 TO 166)*

These pages show the signs seen on the surface of an aerodrome. The signs and pictures are probably more relevant to the Aviation Law syllabus than the Flight Planning syllabus.

TEXT COVERAGE AREAS *(JEPPESEN MANUAL, PAGE 201)*

This page shows the text coverage areas and the abbreviations for the various regions.

APPROACH CHART LEGEND NEW FORMAT *(JEPPESEN MANUAL, PAGES NEW FORMAT 1 TO NEW FORMAT 5)*

This section shows the new format for a briefing strip concept. Again these legend pages will be explained in a later chapter but need to be known.

Chapter 7
Jeppesen Airway Manual - Enroute

INTRODUCTION

The enroute section contains:

- 3 Europe – Low Altitude Enroute Charts
- 1 United States – High Altitude Enroute Chart
- 2 United States – Low Altitude Enroute Charts

This chapter explains the three types of charts.

EUROPE – LOW ALTITUDE ENROUTE CHART

For this explanation, use Chart E (LO) **1**. Open the chart so that you have panel 1 fully open. We will use the route B1 starting at AKELI (N5400 W01000) to illustrate the symbols and meanings on the chart. It might help to have page 57 of the introduction open for this chapter.

The first symbol is a solid triangle (▲) which represents the compulsory reporting point at AKELI.

To the right of the compulsory reporting point is the figure 105°, which represents the track to the next significant position – CONNAUGHT.

About halfway between AKELI and CONNAUGHT is a set of figures:

42
B 1
5000
3900a

42	Represents the mileage between AKELI and CONNAUGHT – 42 nm
B1	Is the airway designator – Bravo 1
5000	Is the Minimum Enroute Altitude (MEA) – 5000 ft
3900a	Is the Route Minimum Off-Route Altitude (Route MORA) – 3900 ft

The next compulsory reporting point is a VOR. Note the large blue arrow pointing to the VOR:

Find the decode for the VOR on Page 53:

CONNAUGHT	The name of the beacon
D	Indicates that a DME capability is available
117.4	The frequency of the VOR
CON	The three letter identifier of the VOR
--·- --- -·	The Morse code for the three letter identifier
N53 54.8 W008 49.1	The latitude and longitude of the beacon

Note that the magnetic track to AKELI is given (285°).

Because this is a low-level chart, the airfield at Connaught is displayed using the civil aerodrome symbol shown on page 55. Around the airfield is a dotted green circle representing a Control Zone (CTR). The limits of the CTR are shown by the figures:

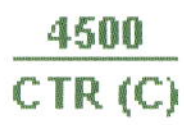

Being a CTR, the lower limit is ground level, the upper limit is 4500 feet. A letter in brackets gives the classification of airspace. In this case (C) represents Class C airspace.

One NDB is located to the southwest of the airfield, to the northeast is a locator beacon.

The frequencies of the beacons are found in the block:

665	The airfield elevation in feet
398 OK	The frequency and two letter identifier of the locator beacon; a green arrow points from this to the beacon.
364 KNK	The frequency and three letter identifier to the south-west.

The symbol that points into the airfield from the east (looking like a paper dart) denotes the direction of the localiser.

Continuing along the airway, the track to the next compulsory reporting point is 113°M. The first symbol along the track:

The total mileage between facilities; in this case CONNAUGHT to DUBLIN. This is based on the distances of 55, 18, and 20.

Written above the track to DUBLIN are the words:

SHANNON CTA **(A)**
FL 200 and below **(C)**

This designates that the airway is within the SHANNON CTA:

- Above FL 200, it is Class A airspace
- FL 200 and below, it is Class C airspace

The next symbol:

Δ — The open triangle represents an On-Request Reporting Point – **RANAR**. Further along the track, another On-Request Reporting Point can be found – **TIMRA**. These reporting points are usually found where airways cross.

At DUBLIN, follow the airway on a track of 130°M following airway B39. **TOLKA** indicates the FIR Boundary (SHANNON to LONDON). The next symbol along the track:

X — Represents an unnamed mileage break, underneath the latitude and longitude is [DUB56].

[DUB56] — This is the NavData Identifier. It is Jeppesen derived and must not be used for any flight planning purposes.

The distance between TOLKA and DUB56 is 21 nm. To the right of the mileage indicator is the symbol:

❺ — This symbol represents additional information or restrictions regarding the airway. If you look approximately 40 nm south of the symbol, you will find a box, labeled “5”, with the information.

The airway is only available 1700 – 0900 LT (Winter) or 1800 – 0800 LT (Summer) and weekends and holidays. This is because it passes through Danger Area EG(D)-202 which is active from MSL to 6000 feet at the times shown. It also passes through EG(D)-PILOTLESS TARGET AREA active from MSL to 60 000 ft.

Following the airway, the next significant symbol is:

The dotted lines form a rough square on the chart. Follow the dotted lines to the northeast corner of the square. The following words appear:

For complete information see
MANCHESTER U.K.
EGCC AREA

The chart is quite difficult to interpret in this area bounded by the dotted lines and so a Manchester Area chart of a larger scale is available.

Enroute Exercise 1

You will require Chart E(LO) **5**.

An aircraft is to fly a route from RAMME (N56 28.7 E008 11.3) to SPIJKERBOOR (N52 32.4 E004 51.2). Route A7 EEL G10 SPY.

Question 1 What is the frequency of RAM?

Question 2 Can an aircraft fly at FL 190 to VESTA, and if not, why not?

Question 3 On the route from VES to TUSKA what do the following represent?

a. **3500T**

b. 

Question 4 What class of airspace is BREMEN?

Question 5 What is the total distance between VES and EEL?

Question 6 At JUIST can an aircraft fly direct to SPY down airway R12?

UNITED STATES – HIGH ALTITUDE ENROUTE CHART

This chart is nearly the same as the low level chart discussed. The front panel includes the effective heights for the routes:

- 18 000 ft to FL 450 in the USA
- 18 000 ft to unlimited in Canada

In Canada RNAV routes are designated with a "T" eg T466. These are effective at and above FL 310. Where an MEA is greater than 18 000 ft, it will be shown on the chart.

All airspace on the chart is Class A from 18 000 ft to 60 000 ft.

Chart coverage is the same as for the low level system.

On the reverse of the front panel:

Communications	The radio frequencies applicable to the ARTCCs that provide coverage over the chart.
Airspace Restricted Areas	The limits and times of operation of the respective areas that are within the chart coverage area. ARTCCs that any area falls in are given.
Cruising Altitudes	Given for both Canada and the USA

No airfields are shown on the chart.

To illustrate that the chart follows the same rules as the low level chart complete the following exercise:

Enroute Exercise 2

You will require Chart US(HI) **3**

An aircraft is to fly the route MIRABEL (N45 53.3 W074 22.6) J546-553 YOW J546 PECK

Question 1	Is there a DME capability at YMX?
Question 2	What is the total distance between YMX and AMERT?
Question 3	To the south-east of YOW is UPLANDS TACAN. Why is a VHF frequency of 108.8 MHz listed?
Question 4	At AMERT, in which direction is the holding pattern?
Question 5	At HEIMS, what is the inbound track of the holding procedure?

UNITED STATES – LOW ALTITUDE ENROUTE CHARTS

Using Chart US(LO) 45/46. The US Low Level Chart front and rear panels are self explanatory. Note that the panel system of finding airfields applies to both charts. The respective panel numbers are shown at the top of the City Location Guide.

If the chart is open, on the opposite side of the front panel are the Part Time Terminal Airspace Hours.

To illustrate that the chart follows the same rules as the low level chart complete the following exercise:

Enroute Exercise 3

You will require Chart US(LO) 46

An aircraft is to fly the route UTICA (Panel 1A) V490 MHT

Question 1 What do the figures 2.1G-2.2-2.65 BUFFALO mean above the listing for UTICA?

Question 2 What is the bearing and distance of PAYGE from ALB?

Question 3 South of GALWA is a CTR for SCHENECTADY. What real-time weather data can you receive and on what frequency?

Question 4 For SCHENECTADY, what does the symbol **(*D)** on the edge of the CTR mean?

Question 5 At CAM what do the letters HIWAS mean?

Question 6 After BRATS the symbol below appears. What does this mean?

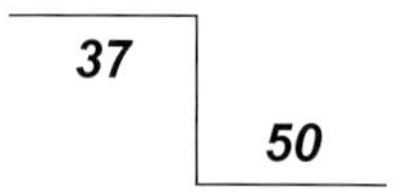

ENROUTE ANSWERS

Enroute Exercise 1

Question 1 112.3 MHz

Question 2 No. E> means that even flight levels must be flown towards VES.

Question 3
a. The minimum obstacle clearance altitude is 3500 feet.
b. Indicates a DME fix

Question 4 Class E

Question 5 157 nm

Question 6 No, as the airway R12 is one way toward JUIST

Enroute Exercise 2

Question 1 Yes

Question 2 166 nm

Question 3 Because civil aircraft can receive DME from a TACAN, the VHF frequency is used to tune the DME which is frequency paired to that respective VHF frequency.

Question 4 Right

Question 5 255°M

Enroute Exercise 3

Question 1 Buffalo guards (receives) on frequency 122.1 MHz and transmits on frequencies 122.2 MHz and 122.65 MHz.

Question 2 322°M/26 nm

Question 3 See the definition of AWOS 3. Altimeter setting, wind data, temperature, dewpoint, density altitude, visibility, cloud, and ceiling data

Question 4 The CTR is Part Time. The Class D airspace is operational 0730LT Monday to 2230LT Friday.

Question 5 Hazardous inflight weather advisory service

Question 6 The navigation frequency changeover point between two stations is indicated by the distance from the station to the point of change. 37 nm to CAM and 50 nm MHT.

Chapter 8
Jeppesen Airway Manual - High

INTRODUCTION

The High section contains:

- 3 Europe – High Altitude Enroute Charts
- 1 Canada/Alaska – High Altitude Enroute Chart
- 1 Atlantic Orientation Chart
- 1 Atlantic – Polar High Altitude Enroute Chart
- 1 North Canada Plotting Chart
- 2 North Atlantic Plotting Chart
- 1 VFR + GPS Chart, Germany ED-6 (Discussed in a later Chapter)

EUROPE – HIGH ALTITUDE ENROUTE CHART

Using Chart E(HI) **1**. The front and rear panels are slightly different from the low level chart. The countries that the chart covers and the upper airspace classification and limits are listed. The area of coverage is the same.

The rear panel covers:

AIRSPACE RESTRICTED AREAS — The area, heights, and times of operation are given

TRANSPONDER SETTINGS — Are listed on pages that are not in the Airway Manual

CRUISING LEVELS — Are found on a panel within the chart

NOTES — Unlike the low chart, the notes are found on this panel. (The other two HI charts follow the same convention as the low level charts.)

Note that there are two E(HI) **4** charts. One is listed **CAA FOR CPL/ATPL EXAMINATIONS**; if this E(HI)4 is to be used, the examination question will say so.

High Exercise 1

Chart E(HI)4

An aircraft is to fly a route from FRANKFURT (N50 03.2 E00838.2) UG1 NTM UJ35 GOROL. Destination EGUM.

Question 1	During what period are weekend routes available for use in France?
Question 2	What are the radio aid facilities at FFM?
Question 3	What is the DME distance of ADENU from NTM?
Question 4	Crossing UJ35 is RUN 852. What does the R suffix mean?
Question 5	The aircraft is flying at FL220. What radar unit and frequency will the aircraft use after Brussels?
Question 6	What frequency does Brussels Weather Use?
Question 7	UG106 is not available. Can the aircraft use UG1, and if so, why?

CANADA/ALASKA – HIGH ALTITUDE ENROUTE CHART CA(HI) 3/4

An explanation of Canadian Airspace is at the top of the front panel. Chart revision and chart coverage are standard. On the reverse panels Cruising Altitudes and Communications are self-explanatory.

On Chart CA(HI) 3, the Gander OCA communications procedures is located in the bottom right hand corner.

For Chart CA(HI) 4, use care when interpreting bearings. As the chart covers areas above 70° North, some bearings are given as true directions. Below 70° North, bearings are given in magnetic.

Other than the differences listed above, the chart is similar to those seen earlier.

ATLANTIC ORIENTATION CHARTS AT(H/L) 1/2

As stated on the front panel, these charts are designed for route planning and over-ocean enroute navigation between the major transatlantic aerodromes.

Chart coverage is standard. A legend for Navaids is listed below the chart coverage.

TRANSPONDER SETTINGS

Details are listed on Chart 2, Panel 1.

CRUISING LEVELS

Cruising levels are shown on the respective enroute charts.

VOLMET BROADCASTS

Selected VOLMET broadcasts are listed. Frequencies are given with the time of operation.

NAVAID INFORMATION

On the reverse of the front panel is a listing of the navaids that feature on the chart.

NORTH ATLANTIC AND CANADA MNPS

Panel 7 on Chart 1 lists the requirements.

NAT ORGANISED TRACK SYSTEM

Panel 8 on Chart 1 lists the requirements.

NORTH ATLANTIC COMMUNICATIONS

Panel 8 on Chart 1 lists the requirements.

NORTH ATLANTIC CROSSING CLEARANCE PROCEDURE AND FREQUENCIES

Panel 9 on Chart 1 lists the requirements. Continuation of any requirements is the on Panel 1 of Chart 2.

POSITION REPORTING PROCEDURES

Panel 1 on Chart 2 lists the requirements.

INCREASED WEATHER REPORTING

Panel 1 on Chart 2 lists the requirements.

SPECIAL PROCEDURES FOR IN-FLIGHT CONTINGENCIES IN MNPS/RVSM AIRSPACE

Panel 1 on Chart 2 lists the requirements.

IN-FLIGHT CONTINGENCY PROCEDURES FOR WAKE VORTEX ENCOUNTERS WITHIN NAT MNPS AIRSPACE

Panel 1 on Chart 2 lists the requirements.

On each chart, information panels give extra information such as:

- Mach Number Technique
- Standard Air-Ground Message Types and Formats
- North Atlantic Communication Equipment Requirement

Before any discussion of the chart is given, complete the following exercise:

High Exercise 2 Chart AT(H/L) **1** and **2**

Question 1 An aircraft is flying a night trip to Rome. To obtain the weather in Rome:

a. Which VOLMET Station would be used?
b. What frequency would be used?
c. At what time past the hour are the weather broadcasts for Rome?

Question 2 For MNPSA operations, two fully serviceable Long Range Navigation Systems (LRNS) are required, a LRNS may be:

a.
b.
c.

Question 3 If both LRNS fail, what are the pilot's actions?

Question 4 Can aircraft without an MNPS capability be cleared to climb/descend through MNPS airspace?

Question 5 To use a westbound flight over NAT OTS, the PRM must be sent no later than?

Question 6 After leaving oceanic airspace what Mach Number must a pilot maintain?

Question 7 How long before departure should an aircraft departing from Prestwick, transiting to oceanic entry point N59 W010 contact ATC for oceanic clearance?

Question 8 Longitude intervals for position reporting at a latitude of 80°N should be?

Question 9 In the case of a radio failure, what is the transponder setting for entering the Bermuda TMA?

DISTANCE

Use Chart AT(H/L) **1**. The chart is a Lambert Conformal projection with a scale of:

AT(H/L) **1**	1 INCH = 132 NM
AT(H/L) **2**	1 INCH = 136 NM

Distance can be measured in three ways:

- By extracting the values that are printed on the published tracks
- By using the scale to the side of the chart
- By using the latitude scale

High Exercise 3 Chart AT(H/L) 1
Use the route ***LASNO*** (N48 35.9 W009 00) to PORTO SANTO

Question 1 The magnetic variation at LASNO is?

Question 2 What is the distance from LASNO to ***ARMED***?

Question 3 What flight level would be flown between FL 280 and FL 310?

Question 4 In which FIR is Porto Santo sited?

Question 5 What navigation aids are available at Porto Santo?

ATLANTIC POLAR HIGH ALTITUDE ENROUTE CHART AT(HI)5

This chart is designed for route planning and high altitude polar navigation between Europe and North America. Where a flight operates between Europe and the Canadian Arctic Control Area between FL 280 and FL 390, it is strongly recommended that Polar Track Structure (PTS) flight planning be used. The PTS system works on promulgating tracks for the following:

Traffic to Alaska	1200 – 1800Z
Traffic to Europe	0000 – 0600Z

The designation PTS is only used when the whole length of a Polar Track is used. If a part route is used it is planned as a random route.

There are:

- 10 fixed tracks in the Reykjavik Control Area
- 5 fixed tracks in the Bodo Oceanic Control Area (these tracks are a continuation of the tracks in the Reykjavik Control Area)

CHART PROJECTION

The following properties apply to a Polar Stereographic Chart:

- Scale is constant and correct
- Great circles are straight lines
- Bearings are correct

On the chart, a grid is superimposed for grid navigation. This grid is aligned with the Greenwich Meridian.

BEACON ALIGNMENT

Some VORs are aligned to true north or grid north. This will be indicated on the chart:

ALERT TACAN	N82 31.0 W062 12.7	Aligned to Grid North
RESOLUTE BAY VOR/DME	N74 43.7 W094 55.4	Aligned to True North

High Exercise 4 Use the route ADN (N57 18.6 W002 15.9) UH70 GONUT PTS2 69W

Question 1 What is the distance from ADN to GONUT?

Question 2 What is the true track from GONUT to N66 00.0 W008 30.0?

Question 3 What is the grid track from N66 00.0 W008 30.0 to GONUT?

PLOTTING ON A POLAR CHART

A grid is superimposed on this chart (aligned to the 0° meridian). This grid is printed because the use of true or magnetic references in Polar regions is difficult because:

- Magnetic variation changes rapidly over short distances
- The magnetic compass becomes unreliable at latitudes greater than 70°N
- The convergence of the meridians causes the course to change rapidly

Please note that other meridians may be used to reference the grid. The same principle applies.

Using the diagram below:

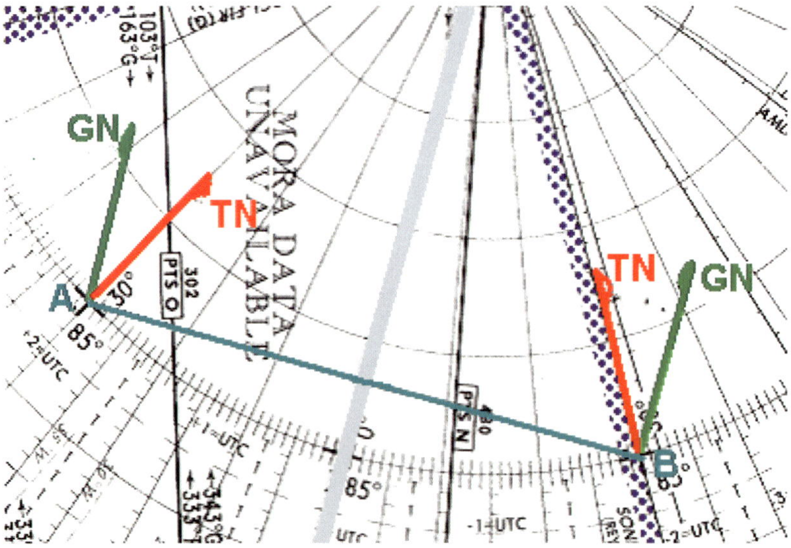

A line is drawn between B (N85 W030) and A (N85 E030).

By inspection, the Grid Course equals the True Course when the line passes through the 0° meridian. Both True North and Grid North are the same.

Grid Course 270°
True Course 270°

However, the true and grid course differs at both A and B.

By measurement if you are transiting from B to A:

At B
Grid Course 270°
True Course 300°

At A
Grid Course 270°
True Course 240°

The angular difference between the two is convergence. If True North is west of Grid North (B) there is a westerly convergence; easterly convergence is where True North is east of Grid North (A).

The angular difference between the Grid North and True North is 30° — the angular difference between the Reference Meridian (0°) and Point A or Point B at 030°.

Following a simple convention:

Convergence west – True best
Point B
Grid Course = True Course -30°

Convergence east – True least
Point A
Grid Course = True Course +30°

If you forget the convention, the formula is written in the bottom right hand corner of the chart.

	+ Longitude West
Grid Bearing = True Bearing	
	- Longitude East

The longitude refers to whether True North is to the west or the east of Grid North.

For Point A, True North is east of Grid North.

Grid Bearing = True Bearing – 30°

High Exercise 5 Use Chart AT(HI) **5**
Route KARLL (N70 W151) to N85 W151

Question 1 What is the grid track to the end point?

Question 2 What is the total route distance?

Question 3 What is the time difference to UTC at 85°N?

Question 4 What is the true bearing of a point N85 W180 from a point N85 W100?

NORTH CANADA PLOTTING CHART (NCP)

A Lambert Conformal Chart with a scale of 1 inch = 120 nm. This chart is designed for plotting routes and position information. Because the chart is a Lambert's Conformal chart, the following apply:

- Scale can be considered constant
- A straight line is a great circle
- Bearings are correct (for a mean course, measure at the centre point)

High Exercise 6 Use Chart **NCP**
Plot the route SHANNON (N52 45 W00855) to GANDER (N49 55 W054 30)

Question 1 What is the distance between Shannon and Gander?

Question 2 What is the mean great circle track?

NORTH ATLANTIC PLOTTING CHART (MAP/NAP)

Exactly the same rules apply to this chart as apply to the NCP.

NORTH ATLANTIC PLOTTING CHART (NAP/INSET)

Both charts have range and time circles for suitable diversion airfields. The airfields are listed by their four-letter ICAO identifier. Circles are provided for:

- 820 NM/120 MIN 410 knots TAS
- 1220 NM/180 MIN 406 knots TAS

EQUAL TIME POINT

Use the **NAP**. For a route from SHANNON (EINN) to GANDER (CYQX). In the centre of the chart is an extended mid-point line. The total range between the two points is 1715 nm. Either side of the central line is a range scale.

This mid-point line can be used in conjunction with the **EQUAL TIME POINT (ETP)** table on the right.

Cruise Level	FL 350
Wind Component Midpoint to Gander	+50 kts
Wind Component Midpoint to Shannon	-70 kts

STEP 1 To calculate the ETP, first calculate an equitime number from the graph.

STEP 2 Enter the graph on the left side with a continuing wind component to Gander **(+50).**

STEP 3 Enter the graph at the bottom with a returning wind component to Shannon **(-70).**

STEP 4 Where the two intersect, read off the nearest equitime number (interpolate to a half number if the intersection is halfway between two lines).

-7

STEP 5 The equitime number is multiplied by 1% of the total distance (1715 nm). This is the distance of the ETP from the midpoint.

-120 nm

The sign is important.

STEP 6 If the product is negative, the ETP is in the returning direction. If the product is positive, the ETP is in the continuing direction.

737 nm from Shannon

High Exercise 7

Route GANDER (CYQX) to KEFLAVIK (BIKF)

Cruise Level	FL 350
Wind Component Midpoint to Keflavik	-70 kts
Wind Component Midpoint to Gander	+100 kts

What is the distance of the ETP from Keflavik?

HIGH EXERCISE ANSWERS

High Exercise 1

Question 1 Friday 1600Z – Monday 0800Z (Front panel of the chart).

Question 2 VOR/DME

Question 3 15 nm

Question 4 RNAV route

Question 5 Brussels Control 127.225 MHz

Question 6 127.80 MHz

Question 7 Yes. Note 4, if the aircraft is below FL 225, the airway is available westbound to EGUM when UG106 is not available.

High Exercise 2

Question 1 An aircraft is flying a night trip to Rome, to obtain the Rome weather.

a. Shannon
b. 3413
c. H+ 20 to 25 minutes or H+ 50 to 55

Question 2 Chart 1, Panel 7

a. One inertial navigation system
b. One OMEGA navigation system
c. One navigation system using the inputs from one or more IRS or OMEGA sensor or any other sensor system complying with MNPS specifications

Question 3 Chart 1, Panel 8

Notify ATC, make the best use of the procedures specified for one system failure, maintain special look out. If no instructions from ATC, consider climbing/descending.

a. 500 ft if below 29 000,
b. 1000 ft if (500 ft if in RVSM) FL 330 to FL 370,
c. if above FL 290 climbing 1000 ft or descending 500 ft if at FL 290

Question 4 Yes (Chart 1, Panel 8)

Question 5 1900 Z (Chart 1, Panel 8)

Question 6 The assigned oceanic mach number unless ATS authorises a change (Chart 1, Panel 8)

Question 7 30 minutes (Chart 1, Panel 9)

Question 8 20° (Chart 2, Panel 1)

Question 9 Mode A/C 2100 (Chart 2, Panel 1)

High Exercise 3

Question 1 8°W

Question 2 448 nm

Question 3 FL 290, note the even arrow on the airway

Question 4 Lisbon

Question 5 VOR/DME

High Exercise 4

Question 1 233 nm

Question 2 342°T

Question 3 168°G

High Exercise 5

Question 1 151°G

Question 2 900 nm

Question 3 +10 hours

Question 4 310°T

High Exercise 6

Question 1 1720 nm

Question 2 263°T

High Level Exercise 7 553 nm

Chapter 9

Jeppesen Airway Manual - ATC, Air Reporting by Voice Communications (AIREP)

AIREP

Use the AIREP form on Page 434A of the Air Traffic Control Section of the Manual. The form is split into three sections:

Section 1

Item 1	Aircraft Identification
Item 2	Position
Item 3	Time
Item 4	Flight Level or Altitude
Item 5	Next Position and Estimated Time
Item 6	Ensuing Significant Point

Section 2

Item 7	ETA
Item 8	Endurance

Section 3

Item 9	Air Temperature
Item 10	Wind Direction
Item 11	Wind Speed
Item 12	Turbulence
Item 13	Aircraft Icing
Item 14	Humidity
Item 15	Phenomenon Prompting a Special Air-Report

ROUTINE AIR REPORTS

For routine air reports, Section 1 is obligatory. You may omit Items 5 and 6 when authorised by Regional Supplementary Procedures. Section 2 is only added when:

- Requested by the operator or the designated representative, or
- When the PIC deems it necessary

Section 3 is added in accordance with Annex 3 and the relevant Regional Supplementary Procedures.

SPECIAL AIR REPORTS

A special air report is made when any of the phenomena under Item 15 are observed. Items 1 to 4 and Item 15 are required from all aircraft. Phenomena listed under SST are only reported by supersonic transport at supersonic and transonic speeds.

Where volcanic activity is reported a post flight report is made on the form found on page 434E – Special Air Report of Volcanic Activity (MODEL VAR).

A Special Air Report is made as soon as is practicable after the phenomenon is observed. If the phenomenon is observed near a time or place that requires a Routine Air Report, a Special Air Report is still made.

REPORTING INSTRUCTIONS

Section 1

Item 1 — Aircraft Identification: Use the aircraft RTF callsign.

Item 2 — Position:

> **Latitude and Longitude** Latitude can be reported as whole degrees (2 characters) or as degrees and minutes (4 characters). Longitude is given as whole degrees (3 characters) or degrees and minutes (5 characters)
>
> 46 North 070 West
>
> 4620 North 07005 West
>
> You can use a significant point as specified by the coded designator
>
> LN (Lima November)
>
> GOROL
>
> Magnetic bearing (3 characters) and distance (2 characters) can also be used
>
> DUB 180 degrees 40 miles

Item 3 — Time: Time is reported in hours and minutes UTC. Where minutes past the hour are reported, use 2 characters . The time used is the actual time at the position, not the time of the report. When making a Special Air Report, time is always given as hours and minutes UTC.

Item 4 — Flight Level or Altitude: Flight level is reported as 3 numerics, such as FL 350. Report altitude in metres or feet. If the aircraft is climbing or descending, report this as well as the level.

Item 5 — Next Position and Estimated Time: The next reporting point and the ETA, or the next estimated position that will be reached one hour later, are reported. Time and position are reported in accordance with Items 2 and 3.

Item 6 — Ensuing Significant Point: Report any significant point after the next reporting point using the same convention as that for Item 5.

Section 2

Item 7 — ETA: Report the name of the airfield of first intended landing with the ETA in hours and minutes UTC.

Item 8 — Endurance: Report endurance in hours and minutes

Section 3

Item 9 — Air Temperature: Report temperature as plus or minus degrees Celsius

Item 10 — Wind Direction and Item 11 — Wind Speed: The spot wind referring to the position in Item 2 is given. Report the wind in °T and the speed in any of the following:

- Knots
- Kilometres per hour
- Metres per second

Item 12 — Turbulence: Report turbulence as:

- Turbulence Severe
- Turbulence Moderate
- Turbulence Light

Severe Turbulence — Conditions in which abrupt changes in aircraft attitude and/or altitude occur. The aircraft may be out of control for short periods. Usually, large variations in airspeed occur.

Changes in accelerometer readings are greater than 1.0g at the aircraft's C of G.

Occupants are forced violently against seat belts. Loose objects are tossed about.

Moderate Turbulence — Conditions in which moderate changes in the aircraft attitude and/or altitude may occur but the aircraft remains in positive control at all times. Usually, small variations in airspeed occur.

Changes in accelerometer readings are 0.5g to 1.0g at the aircraft's C of G.

Occupants feel strain against seat belts. It is difficult to walk. Loose objects move about.

Light Turbulence — Conditions less than moderate turbulence. Changes in accelerometer readings are less than 0.5g at the aircraft's centre of gravity.

Item 13 — Aircraft Icing: Report icing as:

- Icing Light
- Icing Moderate
- Icing Severe

The following appliy:

Severe — Conditions in which immediate change of heading and/or altitude is considered essential

Moderate — Conditions in which a change of heading and/or altitude may be considered desirable

Light — Conditions less than moderate icing

Item 14 — Humidity: Report relative humidity as humidity followed by 3 numerics:
85% RH Humidity 085

Item 15 — Phenomenon Prompting a Special Air-Report: Report one of the following when encountered or observed:

- Turbulence severe
- Icing severe
- Mountain wave severe
  - Conditions in which the accompanying downdraught is 3 mps (600 fpm) or more and/or severe turbulence is encountered
- Thunderstorm
- Thunderstorm with hail
  - Reports are made for thunderstorms that are:
    - Obscured in haze, or
    - Embedded in cloud, or
    - Widespread, or
    - Forming a squall line
- Dust storm or sandstorm heavy
- Volcanic ash cloud
- Pre-eruption volcanic activity or volcanic eruption

Supersonic aircraft at transonic or supersonic speed report the following:

- Turbulence moderate
- Hail
- CB clouds

Information recorded on the MODEL VAR form is not for transmission. On arrival, deliver it to the aerodrome meteorological office. If no office is available then deliver it in accordance with any local arrangements.

Chapter 10
Jeppesen Airway Manual - ATC, The Flight Plan

TYPES AND CATEGORIES OF FLIGHT PLANS

There are two types of flight plans:

- The VFR Plan, and
- The IFR Plan

The flight plans fall into three categories:

Full Flight Plan — Where information is filed on the ICAO flight plan.

Repetitive Flight Plan — Where an operator flies a route on a regular or scheduled basis, then a repetitive flight plan can be filed. These plans are automatically activated at the appropriate time for each flight.

Abbreviated Flight Plan — Where a portion of a route or flight needs to be controlled e.g. flying in a CTR or crossing an airway. These flight plans can be filed by:

- Telephone prior to take-off, or
- R/T when airborne

FILING A FLIGHT PLAN

While a flight plan can be filed for any flight, you must file a flight plan for any of the following flights:

- Any flight, or portion of flight, to be provided with an ATC service
- Any IFR flight in advisory airspace
- Any flight, or within designated areas, or along designated routes, when required by the appropriate ATS authority. This is to facilitate the provision of:
  - Flight information
  - Alerting and search and rescue services
- Any flight across an international boundary

It is advisable to file a VFR or IFR flight plan if the flight involves flying:

- Over the sea more than 10 nm from the coast
- Over a sparsely populated area where SAR would be difficult

SUBMISSION OF A FLIGHT PLAN

Unless otherwise stated, the proper time to file a flight plan for a flight to be provided with an air traffic control service or an air traffic advisory service is:

- 60 minutes before departure (normally this is 60 minutes before the aircraft requests clearance to start up and taxi as the estimated off-block time (EOBT) is used as the planned departure time, not the planned airborne time), or
- If submitted in flight, at a time that ensures its receipt by the appropriate air traffic services unit at least ten minutes before the aircraft is estimated to reach:
  - The intended point of entry into a CTA or advisory area, or
  - The point of crossing an airway or advisory route

For flights subject to ATFM measures or North Atlantic flights, the following procedures apply:

- Flight plans shall be submitted at least three hours before the EOBT
- Any changes to the EOBT of more than 15 minutes shall be the subject of a modification message
- When a repetitive flight plan (RPL) or an individual flight plan (FPL) has been filed but it is decided, within four hours of EOBT, to use an alternative routing between the same aerodromes of departure and destination:
  - A cancellation message (CNL) shall be transmitted immediately to all addressees of the previous flight plan
  - A replacement flight plan (REP) in the form of a FPL with identical callsign shall be transmitted after the CNL message and with a delay of not less than five minutes
  - The last REP shall be filed at least 30 minutes before EOBT

The submission of a REP should be accepted as fulfilling a state's requirement for advance notification of flight.

CONTENTS OF A FLIGHT PLAN

A flight plan is comprised of information that is considered relevant by the appropriate ATS authority:

- Aircraft identification
- Flight rules and type of flight
- Number and types of aircraft and wake turbulence category
- Equipment
- Departure aerodrome
- Estimated off-block time
- Cruising speed(s)
- Cruising level(s)
- Route to be followed
- Destination aerodrome and total elapsed time
- Alternate aerodrome(s)

- Fuel endurance
- Total number of persons on board
- Emergency and survival equipment
- Other information

CHANGES TO A FLIGHT PLAN

Report all changes to a flight plan submitted for an IFR flight, or a VFR flight operated as a controlled flight, to the appropriate ATS unit. For other VFR flights, report significant changes.

CLOSING A FLIGHT PLAN

Make a report of arrival either in person or by radio at the earliest possible moment after landing, to the appropriate ATS unit. This must be done by any flight for which a flight plan has been submitted.

When a flight plan has been submitted for a portion of a flight, it is closed by the appropriate report to the relevant ATS unit.

When no ATS unit exists at an arrival aerodrome, make the arrival report as soon as practicable after landing and by the quickest means available to the nearest ATS unit.

When communication facilities are inadequate and alternative arrangements for the handling of arrival reports on the ground are not available; take the following action:

- Immediately prior to landing, the aircraft transmits by radio to an appropriate ATS unit, a message comparable to an arrival report. This is where a report is required.
- This transmission is made to the aeronautical station serving the ATS unit in charge of the FIR in which the aircraft is operating

Arrival reports made by aircraft contain the following elements:

- Aircraft identification
- Departure aerodrome
- Destination aerodrome (only in the case of a diversionary landing)
- Arrival aerodrome
- Time of arrival

USE OF REPETITIVE FLIGHT PLANS (RPLs)

General RPLs are not to be used for flights other than:

- IFR flights operated regularly on:
  - The same day(s) of consecutive weeks, and
  - On at least ten occasions, or
  - Every day over a period of at least 10 consecutive days. The elements of each flight plan shall have a high degree of stability.

RPLs cover the entire flight from the departure aerodrome to the destination aerodrome. RPL procedures are only applied when all ATS authorities concerned with the flights have agreed to accept RPLs.

The use by States of RPLs for international flight are subject to the provision that the affected adjacent States either already use RPLs or will use them at the same time. The procedures for use between States are subject to bilateral, multilateral, or regional air navigation agreement as appropriate.

CHANGE FROM IFR TO VFR FLIGHT

A change from IFR fight to VFR flight is only acceptable when a message is initiated by the PIC containing the specific expression *"Cancelling my IFR flight".* No invitation to change from IFR flight to VFR flight is to be made either directly or by inference.

No reply, other than the acknowledgement *"IFR flight cancelled at ... (time)",* should normally be made by an ATS unit.

When an ATS unit possesses information that IMC are likely to be encountered along the route of flight, a pilot changing from IFR flight to VFR flight should, if practicable, be so advised.

An ATS unit receiving notification of an aircraft's intention to change from IFR flight to VFR flight shall as soon as practicable inform all other ATS units to whom the IFR flight plan was addressed, except those units the flight has already passed.

ADHERENCE TO FLIGHT PLAN

Except where stated, an aircraft adheres to the current flight plan or the applicable portion of a current flight plan submitted for a controlled flight, unless:

- Requests for a change to a flight plan have been made to the appropriate ATC unit; clearance must be obtained before any changes can be made, or
- An emergency situation arises which necessitates immediate action by the aircraft. In such an event, as soon as circumstances permit after emergency authority is exercised, notify the appropriate ATS unit of the action taken.

Unless otherwise authorized or directed by the appropriate ATC unit, controlled flights:

- When on an established ATS route, operate along the defined centre line of that route, or
- When on any other route, operate directly between the navigational facilities and/or points defining that route

Aircraft operating along an ATS route segment defined by reference to VOR should change over navigation guidance from the facility behind the aircraft to that ahead of it at, or as close as operationally feasible to, the change over point.

Any deviation from the above requirements is notified to the appropriate ATS unit.

INADVERTENT CHANGES

In the event that a controlled flight inadvertently deviates from its current flight plan; the following action is taken:

Deviation From Track — If the aircraft is off-track, action shall be taken forthwith to adjust the heading of the aircraft to regain track as soon as practicable.

Variation in TAS — If the average TAS at cruising level between reporting points varies, or is expected to vary, by ± 5% of the true airspeed from that given in the flight plan, the appropriate ATS unit shall be informed.

Change in Estimate Time — If the time estimate for the next applicable reporting point, FIR boundary, or destination aerodrome, whichever comes first, is found to be in error in excess of ± 3 minutes from that notified to ATS, or such other period of time as is prescribed by the appropriate ATS authority or on the basis of air navigational regional agreements, the appropriate ATS unit shall be notified with a revised estimate time as soon as possible.

INTENDED CHANGES

Requests for flight plan changes include the following:

Change of Cruising Level

- Aircraft identification
- Requested new cruising level and cruising speed at this level
- Revised time estimates (when applicable) at subsequent FIR boundaries

Change of Route

Destination Unchanged

- Aircraft identification
- Flight rules
- Description of new route of flight including related flight plan data beginning with the position from which requested change of route is to commence
- Revised time estimates
- Any other pertinent information

Destination Changed

- Aircraft identification
- Flight rules
- Description of revised route of flight to revised destination aerodrome including related flight plan data, beginning with the position from which the requested change of route is to commence
- Revised time estimates
- Alternate aerodrome(s)
- Any other pertinent information

WEATHER DETERIORATION BELOW THE VMC

When it becomes evident that flight in VMC in accordance with the current flight plan is not practicable, a VFR flight operated as a controlled flight will:

- Request an amended clearance enabling the aircraft to continue in VMC to destination or to an alternate aerodrome, or to leave the airspace within which an ATC clearance is required, or
- If no clearance can be obtained, continue to operate in VMC and notify the appropriate ATC unit of the action being taken either to leave the airspace concerned or to land at the nearest suitable aerodrome, or
- If operated within a CTR, request authorization to operate as a Special VFR flight, or
- Request clearance to operate in accordance with the IFR.

DATE OF FLIGHT IN A FLIGHT PLAN

PANS-RAC states that "if a flight plan is filed more than 24 hours in advance of the EOBT of the flight to which it refers, that flight plan shall be held in abeyance until at most 24 hours before the flight begins so as to avoid the need for the insertion of a date group into that flight plan." The following removes this restriction and specifies details regarding the optional insertion of a date group into the flight plan.

If a flight plan for a flight conducted wholly in the EUR Region is filed more than 24 hours in advance of the estimated EOBT, it is mandatory to provide the date of the flight. If the flight plan is filed less than 24 hours in advance of the EOBT, the date of the flight may be optionally indicated. This information will be inserted in Item 18 of the flight plan in the form of a three-letter indicator followed by an oblique stroke and date of flight in a six-figure format (This is described later in this chapter).

e.g. DOF/YYMMDD

DOF	Date of flight
YY	Year
MM	Month
DD	Day

These flight plans shall be processed and transmitted without being held in abeyance.

COMPLETION OF THE ICAO FLIGHT PLAN

For this part of the chapter you need to refer to the Jeppesen Airway Manual, Air Traffic Control, Page 434H. Certain general rules must be observed when the flight plan is filled in:

- Use upper case at all times
- Adhere to the prescribed format
- Complete all items in accordance with the following instructions
- Fill in data by starting in the first space provided
- Do not insert spaces or obliques where they are not required
- Time should be inserted as a four-figure UTC
- Elapsed times are inserted in four figures, hours and minutes

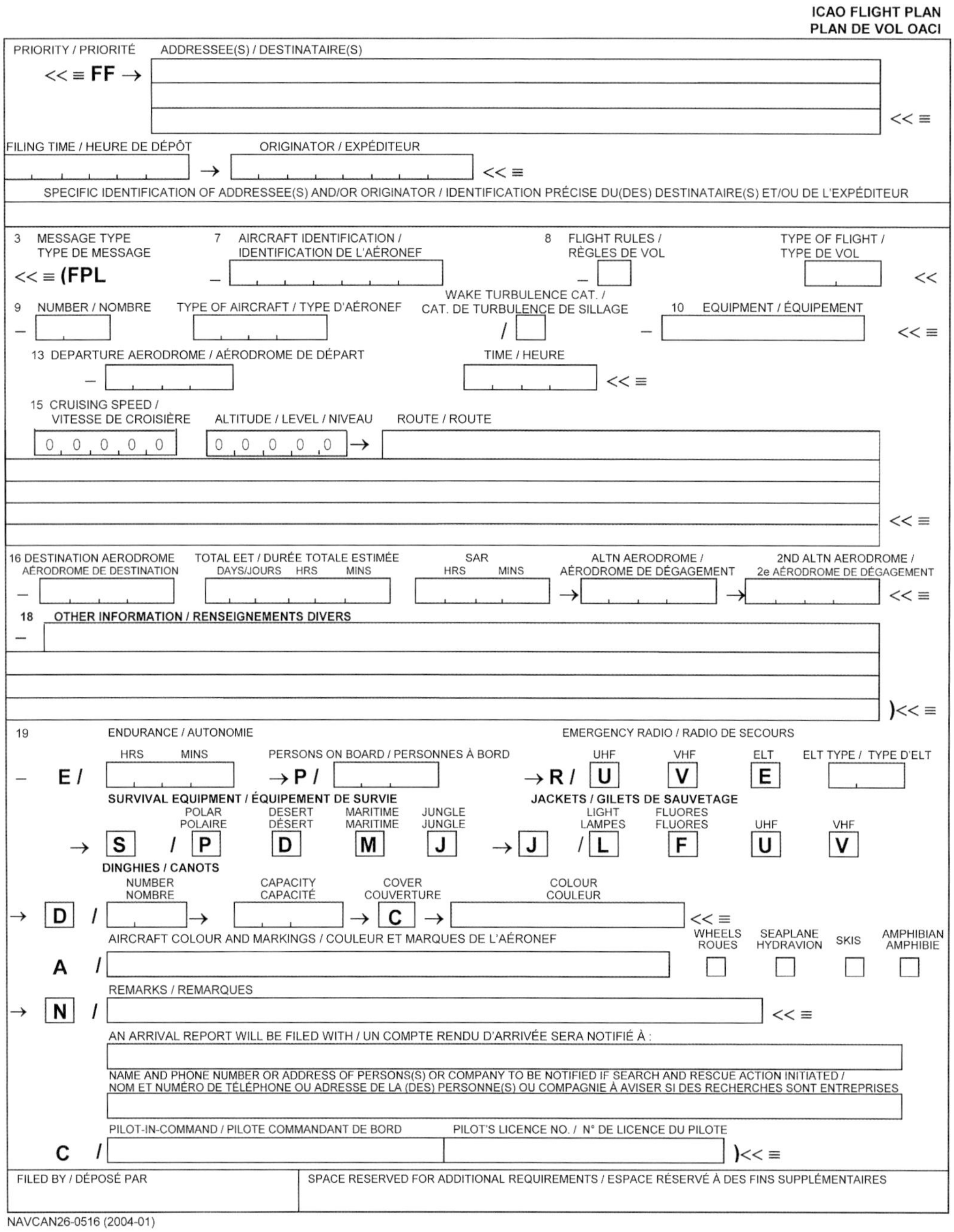

ICAO FLIGHT PLAN
PLAN DE VOL OACI

PRIORITY / PRIORITÉ ADDRESSEE(S) / DESTINATAIRE(S)
<< ≡ FF →
<< ≡
FILING TIME / HEURE DE DÉPÔT ORIGINATOR / EXPÉDITEUR
→ << ≡
SPECIFIC IDENTIFICATION OF ADDRESSEE(S) AND/OR ORIGINATOR / IDENTIFICATION PRÉCISE DU(DES) DESTINATAIRE(S) ET/OU DE L'EXPÉDITEUR

3 MESSAGE TYPE / TYPE DE MESSAGE
<< ≡ (FPL
7 AIRCRAFT IDENTIFICATION / IDENTIFICATION DE L'AÉRONEF
8 FLIGHT RULES / RÈGLES DE VOL
TYPE OF FLIGHT / TYPE DE VOL <<
9 NUMBER / NOMBRE
TYPE OF AIRCRAFT / TYPE D'AÉRONEF
WAKE TURBULENCE CAT. / CAT. DE TURBULENCE DE SILLAGE
10 EQUIPMENT / ÉQUIPEMENT << ≡
13 DEPARTURE AERODROME / AÉRODROME DE DÉPART
TIME / HEURE << ≡
15 CRUISING SPEED / VITESSE DE CROISIÈRE
ALTITUDE / LEVEL / NIVEAU
ROUTE / ROUTE
0 0 0 0 0 | 0 0 0 0 0 →
<< ≡
16 DESTINATION AERODROME / AÉRODROME DE DESTINATION
TOTAL EET / DURÉE TOTALE ESTIMÉE DAYS/JOURS HRS MINS
SAR HRS MINS
ALTN AERODROME / AÉRODROME DE DÉGAGEMENT
2ND ALTN AERODROME / 2e AÉRODROME DE DÉGAGEMENT << ≡
18 OTHER INFORMATION / RENSEIGNEMENTS DIVERS
)<< ≡
19 ENDURANCE / AUTONOMIE HRS MINS
PERSONS ON BOARD / PERSONNES À BORD
EMERGENCY RADIO / RADIO DE SECOURS UHF VHF ELT ELT TYPE / TYPE D'ELT
– E / → P / → R / U V E
SURVIVAL EQUIPMENT / ÉQUIPEMENT DE SURVIE
POLAR / POLAIRE DESERT / DÉSERT MARITIME / MARITIME JUNGLE / JUNGLE
JACKETS / GILETS DE SAUVETAGE
LIGHT / LAMPES FLUORES / FLUORES UHF VHF
→ S / P D M J → J / L F U V
DINGHIES / CANOTS
NUMBER / NOMBRE CAPACITY / CAPACITÉ COVER / COUVERTURE COLOUR / COULEUR
→ D / → → C → << ≡
AIRCRAFT COLOUR AND MARKINGS / COULEUR ET MARQUES DE L'AÉRONEF
WHEELS / ROUES SEAPLANE / HYDRAVION SKIS AMPHIBIAN / AMPHIBIE
A /
REMARKS / REMARQUES
→ N / << ≡
AN ARRIVAL REPORT WILL BE FILED WITH / UN COMPTE RENDU D'ARRIVÉE SERA NOTIFIÉ À :
NAME AND PHONE NUMBER OR ADDRESS OF PERSONS(S) OR COMPANY TO BE NOTIFIED IF SEARCH AND RESCUE ACTION INITIATED / NOM ET NUMÉRO DE TÉLÉPHONE OU ADRESSE DE LA (DES) PERSONNE(S) OU COMPAGNIE À AVISER SI DES RECHERCHES SONT ENTREPRISES
PILOT-IN-COMMAND / PILOTE COMMANDANT DE BORD
PILOT'S LICENCE NO. / N° DE LICENCE DU PILOTE
C /)<< ≡
FILED BY / DÉPOSÉ PAR
SPACE RESERVED FOR ADDITIONAL REQUIREMENTS / ESPACE RÉSERVÉ À DES FINS SUPPLÉMENTAIRES

NAVCAN26-0516 (2004-01)

ITEM 3 – MESSAGE TYPE

You do not fill in this space.

ITEM 7 – AIRCRAFT IDENTIFICATION

A maximum of 7 characters can be inserted in this field.

ITEM 7 — 7 AIRCRAFT IDENTIFICATION / IDENTIFICATION DE L'AÉRONEF

The aircraft identification can be entered in three ways not exceeding the seven characters available:

- The registration mark of the aircraft e.g. GBOBA, N2345AA, OOBAD, when
  - The callsign used is the same e.g. GBOBA or it is preceded by the ICAO telephony designator for the aircraft operating agency e.g. SABENA OOBAD
  - The aircraft is not equipped with radio
- The ICAO designator for the aircraft operating agency followed by the flight identification e.g. AAG234, BAW278 when the callsign to be used by the aircraft consists of the ICAO telephony designator for the operating agency followed by the flight identification e.g. Atlantic 234, Speedbird 278
- The callsign determined by military authorities if this is how the aircraft is identified during flight

ITEM 8 – FLIGHT RULES AND TYPE OF FLIGHT

ITEM 8 — 8 FLIGHT RULES / RÈGLES DE VOL — TYPE OF FLIGHT / TYPE DE VOL

Flight Rules — One of the following letters is used to denote the category of flight rules with which the pilot intends to comply:

I if IFR
V if VFR
Y if IFR first
Z if VFR first

If Y and Z are used, the point or points where a change of flight rules is planned is inserted in Item 15.

Type of Flight — The type of flight is designated by one of the following letters:

S if scheduled air service
N if non-scheduled air transport operation
G if general aviation
M if military
X if a flight other than the ones defined above

ITEM 9 – NUMBER OF AIRCRAFT, TYPE OF AIRCRAFT, WAKE TURBULENCE CATEGORY

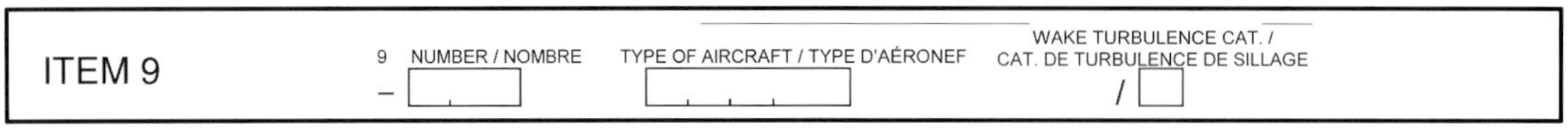

Number of Aircraft — The number of aircraft is entered only if there is more than one e.g. 03

Type of Aircraft — The appropriate designator as specified in ICAO DOC 8643 – Aircraft Type Designators e.g. PA28 (2 to 4 characters may be used)

If no designator has been assigned, or in the case of a formation flight where more than one type of aircraft is being used, insert the ZZZZ. In Item 18, enter the numbers and type of aircraft using the prefix TYP/.

Wake Turbulence Category — Use one of the following letters:

H	HEAVY	To indicate an aircraft with a maximum certificated take-off weight of 136 000 kg or more
M	MEDIUM	To indicate an aircraft with a maximum certificated take-off weight of less than 136 000 kg but more than 7000 kg
L	LIGHT	To indicate an aircraft with a maximum certificated take-off weight of 7000 kg or less

ITEM 10 – RADIO COMMUNICATION, NAVIGATION AND APPROACH AID EQUIPMENT

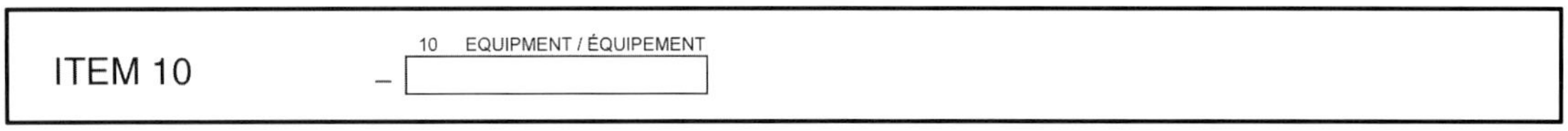

Insert one letter as follows:

N If no COM/NAV Approach aid equipment for the route to be flown is carried, or the equipment is unserviceable, or

S If the standard or prescribed (e.g. NAT requirements) COM/NAV/Approach aid equipment for the route to be flown is carried and serviceable. Unless another combination is prescribed by the appropriate ATS authority, standard equipment is considered to be:

VHF RTF
ADF
VOR
ILS

And/or

Insert one or more of the following letters to indicate the COM/NAV/Approach aid equipment available and serviceable:

Letter	Allocation	Letter	Allocation
A	Not allocated	M	Omega
B	Not allocated	O	VOR
C	Loran C	P	Not allocated
D	DME	Q	Not allocated
E	Not allocated	R	RNP type certification Inclusion of letter **R** indicates that an aircraft meets the RNP type prescribed for the route segment(s), route(s) and/or area concerned
F	ADF	T	TACAN
G	GNSS	U	UHF RTF
H	HF RTF	V	VHF RTF
I	Inertial Navigation	W	RVSM approved If a flight is approved to operate at RVSM levels, the letter **W** must be included.
J	Data Link If the letter J is used, specify in Item 18 the equipment carried, preceded by DAT/ followed by one or more letters as appropriate.	X	MNPS approved In order to signify that a flight is approved to operate in NAT MNPS Airspace, the letter **X** must be inserted, in addition to the letter **S**.
K	MLS	Y	Radio with 8.33 KHz spacing All aircraft which are carrying 8.33 KHz spacing capable radio equipment must insert **Y** into Item 10.
L	ILS	Z	Other equipment If the letter Z is used, specify in Item 18 the other equipment carried, preceded by COM/ and/or NAV/ as appropriate.

Surveillance Equipment — After the oblique stroke in Item 10, insert one or two of the following to describe the serviceable surveillance equipment carried:

SSR Equipment

N	Nil			
A	Transponder	Mode A	4096 Codes	
C	Transponder	Mode A	4096 Codes	Mode C
X	Transponder	Mode S	Without pressure altitude and without aircraft identification transmission	
P	Transponder	Mode S	With pressure altitude but without aircraft identification transmission	
I	Transponder	Mode S	Without pressure altitude but with aircraft identification transmission	
S	Transponder	Mode S	With both pressure altitude and aircraft identification transmission	

ADS Equipment

D ADS capability

ITEM 13 – DEPARTURE AERODROME, AND TIME

ITEM 13	13 DEPARTURE AERODROME / AÉRODROME DE DÉPART –	TIME / HEURE

Departure Aerodrome — Insert the four letter ICAO location indicator of the aerodrome of departure.

If no location identifier has been assigned, insert **ZZZZ** and specify in Item 18 the name of the aerodrome, preceded by DEP/.

If the flight plan is received from an aircraft in flight, insert **AFIL** and specify in item 18 the four letter ICAO location indicator of the location of the ATS unit from which supplementary flight plan data can be obtained, preceded by DEP/.

Time — For a flight plan submitted before departure, insert the estimated off-block time (EOBT) using four characters.

For a flight plan received from an aircraft in flight, use the actual or estimated time over the first point of the route to which the flight plan applies.

ITEM 15 – CRUISING SPEED, LEVEL, AND ROUTE

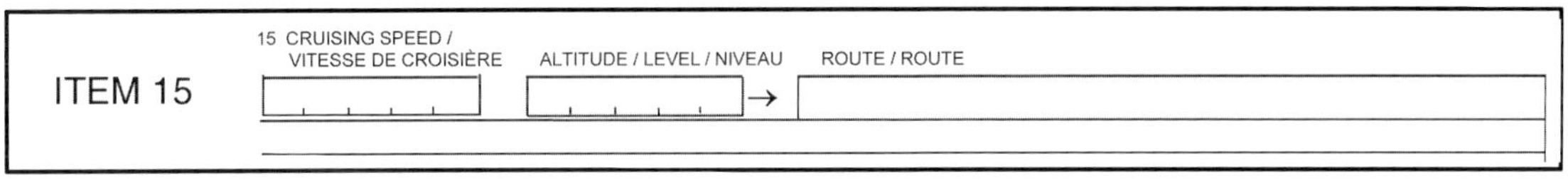
ITEM 15
15 CRUISING SPEED / VITESSE DE CROISIÈRE
ALTITUDE / LEVEL / NIVEAU
ROUTE / ROUTE

Cruising Speed — Insert the cruising speed (TAS) in accordance with the rules shown below:

Knots	Expressed as **N** followed by four figures	N0485
Kilometres Per Hour	Expressed as **K** followed by four figures	K0850
Mach Number	When prescribed by the appropriate ATS authority to the nearest hundredths of unit mach, expressed as **M** followed by three figures M085	

Cruising Level — Insert the planned cruising level for the first or the whole portion of the route to be flown in any of the following formats:

Flight Level	Expressed as **F** followed by three figures	F085, F330
Altitude in hundreds of feet	Expressed as **A** followed by three figures	A045, A100
Standard Metric Level in tens of metres	Expressed as **S** followed by four figures When so prescribed by the appropriate ATS authorities	S1130
Altitude in tens of metres	Expressed as M followed by four figures When so prescribed by the appropriate ATS authorities	M0840
VFR Flights	Where the flight is not planned to be flown at a specific cruising level the letters **VFR**	

Route (Including changes of speed, level, and/or flight rules)

Using the conventions listed below, separating each sub-item by a space, enter the relevant route as detailed in the route requirements section.

ATS Route (2 to 7 characters)

The coded designator assigned to the route segment including, where appropriate, the coded designator assigned to the standard departure or arrival route.

e.g. BCN1, B1, R14, UB10, KODAP2A

Significant Point (2 to 11 characters)
The coded designator (2 to 5 characters) assigned to the point.
e.g. LN, MAY, HADDY

If no coded designator has been assigned, one of the following ways:

Degrees Only (7 characters)
Two figures describing latitude in degrees, followed by N (North) or S (South), followed by three figures describing longitude in degrees followed by E (East) or W (West). The figures should be made up by the insertion of zeros where appropriate.
e.g. 46N078W

Degrees and Minutes (11 Characters)
Four figures describing latitude in degrees and tens and units of minutes, followed by N (North) or S (South), followed by three figures describing longitude in degrees and tens and units of minutes followed by E (East) or W (West). The figures should be made up by the insertion of zeros where appropriate.
e.g. 4620N07805W

Bearing and Distance from a Navigation Aid
The identification of the navigation aid (normally a VOR) in the form of 2 or 3 characters, then the bearing from the aid in the form of three figures giving degrees magnetic, then the distance from the aid in the form of three figures expressing nautical miles. Make up the correct number of figures where necessary by the insertion of zeros.
e.g. To express a point 160°M at a distance of 80 nm from VOR POL
POL160080

Change of Speed or Level (Maximum 21 characters)
The point at which a change of speed (5% TAS or 0.01 M or more) or a change of level is planned is expressed exactly as above followed by an oblique stroke then the cruising speed and the cruising level without a space between them.

e.g.	LN/N0284	LN/N0284A045
	MAY/F180	MAY/N0350F180
	HADDY/N0420	HADDY/N0420F330
	4620N07805W/N0500	4620N07805W/N0500F350
	46N078W/F330	46N078W/M082F330
	POL180080/N0305	POL180080/N0305M0840

Change of Flight Rules (Maximum of 3 Characters)
The point at which the change of flight rules is planned is expressed exactly as above as appropriate followed by a space and one of the following:

VFR	When changing from IFR to VFR
IFR	When changing from VFR to IFR

e.g.	LN VFR	LN/N0284A045 VFR
	MAY IFR	MAY/N0350F180 IFR

Cruise Climb (Maximum 28 Characters)

The letter **C** followed by an oblique stroke, then the point at which the cruise climb is planned to start, expressed in exactly the same way as above. This is followed by an oblique stroke, then the speed to be maintained during the cruise climb followed by the two levels defining the layer to be occupied during the cruise climb or the level above which the cruise climb is planned followed by the letters PLUS.

e.g. C/48N050W/M082F290F350
C/48N050W/M082F290PLUS

ROUTE REQUIREMENTS - GENERAL

Requirements for Flight Along Designated Routes

If the departure aerodrome is located on, or is connected to, the ATS route, insert the designator of the first ATS route.

15 CRUISING SPEED / VITESSE DE CROISIÈRE	ALTITUDE / LEVEL / NIVEAU		ROUTE / ROUTE
N0450	F370	→	UA1 RBT UG32 TOP UB25 BEROK UA41 TAQ DCT

If the departure aerodrome is not on, or is not connected to, the ATS route, use the letters **DCT** followed by the point of joining the first ATS route, followed by the designator of the ATS route.

15 CRUISING SPEED / VITESSE DE CROISIÈRE	ALTITUDE / LEVEL / NIVEAU		ROUTE / ROUTE
N0450	F370	→	**DCT** RBT UG32 TOP UB25 BEROK UA41 TAQ DCT

Then, as shown in the two diagrams above, insert each point at which either a change of speed or level, a change of ATS route and/or a change of flight rules is planned

This is followed in each case by the designator of the next ATS route segment, even if the same as the previous one. **DCT** is used if the flight to the next point is outside a designated route, unless geographical co-ordinates define both points.

Where transition is planned between a lower and upper ATS route and the routes are orientated in the same direction, the point of transition need not be inserted.

Requirements for Flights Outside designated ATS Routes

Points not normally more than 30 minutes flying time or 200 nm apart are inserted, including each point at which a change of speed or level, or a change of flight rules is planned.

When required by the appropriate ATS authority, define:

- The tracks of flights operating predominantly in an east-west direction between 70°N and 70°S by reference to significant points formed by the intersection of half or whole degrees of latitude with meridians spaced at intervals of 10° of longitude.
- The tracks of flights operating in areas outside the above latitudes by significant points formed by the intersection of parallels of latitude with meridians normally spaced at 20° of longitude.
- The tracks of flights operating predominantly in a north-south direction, by reference to significant points formed by the intersection of whole degrees of longitude with specified parallels of latitude, which are spaced at 5°.

The distance between significant points shall, as far as possible, not exceed one hour's flight time. Additional significant points shall be established as deemed necessary.

Insert **DCT** between successive points unless both points are defined by geographical co-ordinates or by bearing and distance.

NORTH ATLANTIC (NAT) FLIGHTS

Requirements for Flight Plans on Random Route Segments at/or South of 70°N

Turbo-jet aircraft should indicate their proposed speeds in the following sequence:

- Cruising Speed (TAS) in knots
- Oceanic entry point and cruising Mach Number
- Landfall fix and cruising speed (TAS) in knots

For all other aircraft the speed is given as TAS.

The flight level for oceanic entry must be inserted at either:

- The last domestic reporting point prior to ocean entry, or
- When at the Oceanic Control Area (OCA) boundary

The route of the flight should be inserted in terms of the following significant points:

- The last domestic reporting point prior to ocean entry
- OCA boundary entry point
  - Required by the Shanwick, New York, and Santa Maria Oceanic Area Control Centres (OACs)
- Significant points formed by the intersection of half or whole degrees of latitude with meridians spaced at intervals of 10° from the Greenwich Meridian to longitude 070°W

  The distance between points shall, as far as possible, not exceed one hour's flight time.

- OCA boundary exit point
  - Required by the Shanwick, New York, and Santa Maria Oceanic Area Control Centres
- The first domestic reporting point after the ocean exit

15 CRUISING SPEED / VITESSE DE CROISIÈRE | ALTITUDE / LEVEL / NIVEAU | ROUTE / ROUTE

N0481 | F310 | → UGI STU UN546 DEVOL UN546
MASIT/M084F310 56N020W 57N030W 56N040W 54N050W
CARPE REDBY/N0480F350 NI86 TOPPS TRAIT/N0441F240

Requirements for Flight Plans on Organised Track System (OTS) South of 70°N

Insert the speed in terms of Mach Number. Also insert the flight level at the commencement point of the OTS.

15 CRUISING SPEED / VITESSE DE CROISIÈRE | ALTITUDE / LEVEL / NIVEAU | ROUTE / ROUTE

N0493 | F280 | → UR14 STU UGI SHA UN535 BURAK
UN533 54N015W/M082F310 NATE YAY/NO487F310 N168
TOPPS/N0481F350 N168 AEX

If the flight is planned to operate along the whole length of one of the organised tracks as detailed in the NAT track message, use the abbreviation "NAT" followed by the code letter assigned to the track without any spacing, as shown above.

Flights wishing to join or leave an organised track at some intermediate point are considered random route aircraft and full route details must be specified in the flight plan. The track letter should not be used to abbreviate any portion of the route in these circumstances.

Each point at which the change in speed or level is requested must be specified as geographical co-ordinates in latitude and longitude, or as a named waypoint.

Requirements for Flight Plans on Random Route Segments North of 70°N

Turbo-jet aircraft should indicate their proposed speeds in the following sequence:

- Cruising Speed (TAS) in knots
- Oceanic entry point and cruising Mach Number
- Landfall fix and cruising speed (TAS) in knots

For all other aircraft the speed is given as TAS.

The flight level for oceanic entry must be inserted at either:

- The last domestic reporting point prior to ocean entry, or
- When at the Oceanic Control Area (OCA) boundary

The route of the flight should be inserted in terms of the following significant points:

- The last domestic reporting point prior to ocean entry
- OCA boundary entry point
  - Required by the Shanwick, New York, and Santa Maria Oceanic Area Control Centres (OACs)
- Significant points formed by the intersection of half or whole degrees of latitude with meridians spaced at intervals of 20° from the Greenwich Meridian to longitude 060°W.

  The distance between points shall, as far as possible, not exceed one hour's flight time.

- OCA boundary exit point
  - Required by the Shanwick, New York, and Santa Maria Oceanic Area Control Centres
- The first domestic reporting point after the ocean exit

Each point at which the change in speed or level is requested must be specified and followed, in each case, by the next significant point.

Requirements for Flight Plans on Polar Track Structure (PTS)

Insert speed in terms of Mach Number at the commencement point of the PTS or at the NAT OCA boundary. Also insert the flight level at the commencement point of the PTS or at the NAT OCA boundary.

Insert the abbreviation "PTS," followed by the code assigned to the track without a space, if the flight is planned to operate along the whole length of one of the Polar Tracks.

15 CRUISING SPEED / VITESSE DE CROISIÈRE — ALTITUDE / LEVEL / NIVEAU — ROUTE / ROUTE

N0489 F310 → UA2 POL UB4 TLA DCT WIK/N0488F350
DCT LIRKI/M084F350 PTSQ LT DCT 8017N11500W DCT
TAYTA M452 HARVZ DCT ENN J125 TAGER

Flights wishing to join or leave a Polar Track at some intermediate point are considered random route aircraft and the full track details must be specified in the flight plan. Do not use the track number to abbreviate any portion of the route in these circumstances.

Each point at which a change in speed or level is requested must be specified as geographical co-ordinates in latitude and longitude followed in each case by the abbreviation "PTS" and the track code.

Requirements for Flight Plans predominantly North/South or South/North The speed is inserted:

- For turbojets in terms of Mach Number
- All other aircraft in terms of TAS

In both cases, the speed is to be specified at either the last domestic reporting point prior to ocean entry or the OCA boundary.

Inserting the route is described in terms of the following:

- The last domestic reporting point prior to ocean entry
- OCA boundary entry point
  - Required by the Shanwick, New York, and Santa Maria Oceanic Area Control Centres (OACs)
- Significant points formed by the intersection of whole degrees of longitude with specified parallels of latitude, which are spaced at intervals of 5° from 20°N to 90°N.

  The distance between points shall, as far as possible, not exceed one hour's flight time.

- OCA boundary exit point
  - Required by the Shanwick, New York, and Santa Maria Oceanic Area Control Centres
- The first domestic reporting point after the ocean exit

Each point at which the change in speed or level is requested must be specified and followed, in each case, by the next significant point.

Requirements For Flight Plans on NAM/CAR Route Structure

The speed is inserted:

- For turbojets in terms of Mach Number
- All other aircraft in terms of TAS

In both cases, specify the speed at the commencement point of the NAM/CAR route structure.

The flight level for the ocean entry must be inserted at the commencement point of the NAM/CAR route structure.

The route of the flight is described in terms of the NAM/CAR ATS route identifiers.

Each point at which either a change in speed or level is requested must be specified and followed in each case by the next route segment expressed by the appropriate ATS route identifier(s), or as a named waypoint.

ITEM 16 – DESTINATION AERODROME, TOTAL ELAPSED TIME, AND ALTERNATE AERODROMES

ITEM 16	16 DESTINATION AERODROME / AÉRODROME DE DESTINATION	TOTAL EET / DURÉE TOTALE ESTIMÉE DAYS/JOURS HRS MINS	SAR HRS MINS	ALTN AERODROME / AÉRODROME DE DÉGAGEMENT	2ND ALTN AERODROME / 2e AÉRODROME DE DÉGAGEMENT
	–			→	→

Destination Aerodrome — Insert the ICAO four-letter location indicator. If no location indicator has been assigned, then the procedure followed is the same for the departure aerodrome except that in Item 18 the name of the aerodrome is preceded by DEST/.

Total Elapsed Time — For IFR flights, this is the total estimated time required from take-off until arriving over:

- ➢ The designated point from which it is intended that an Instrument Approach Procedure, defined by reference to navigation aids, will be commenced, or
- ➢ If no navigation aid is associated with the destination aerodrome, until arriving over the destination aerodrome itself.

For VFR flights it is the estimated total time required from take-off until arriving over the destination aerodrome.

For a flight plan received from an aircraft in flight, the total estimated elapsed time is the estimated time from the first point of the route to which the flight plan applies.

Alternate Aerodromes — Insert the ICAO four letter location indicator. If no location indicator is assigned, then follow the same procedure as for the departure aerodrome except that in Item 18, precede the name of the aerodrome by ALTN/. Only specify two alternate aerodromes may be specified.

ITEM 18 – OTHER INFORMATION

If there is no "other information" a **0** is entered. For NAT flights EET, REG, and SEL should always be included in the sequence.

Where any other necessary information is entered or required, insert it in the following preferred sequence:

REP/ For use by flights in the EUR Region on routes subject to Air Traffic Flow Management to identify a Replacement Flight Plan. After the oblique stroke insert Qn where n represents the sequence number of the Replacement Flight Plan.

e.g. RFP/Q1

EET/ Followed by significant waypoints or FIR boundary designators plus Accumulated estimated elapsed times from take-off to such points where prescribed by Regional Navigation Agreements, or by the appropriate ATS Authority.

e.g. EET/EISN0035
EET/90W0200

NAT Requirements For flights conducted in the NAT Region on random routes, accumulated estimated elapsed times are required for:

- The last domestic reporting point prior to ocean entry
- The OCA boundary entry point
- Each significant point described in Item 15
- The OCA boundary exit point
- The first reporting point on the domestic track

For flights operating along the entire length of an OTS, estimated elapsed times are required for the commencement point of the track and FIR Boundaries.

For flights operating along the whole length of one of the PTS tracks, accumulated estimated elapsed times are required for the commencement point and for each significant point of the track thereafter.

Shanwick, New York, and Santa Maria OCAs require elapsed times to the OCA boundaries only.

RIF/ The route details to the revised destination aerodrome, followed by the ICAO four letter location indicator of the aerodrome. The revised route is subject to re-clearance in flight.

e.g. RIF/DTA HEC HECA
RIF/LEMD

REG/ The registration markings of the aircraft, if different from the aircraft Identification in Item 7. Aircraft registration should be assigned to this field for MNPS flights.

SEL/ SELCAL Code, if so prescribed by the appropriate ATS authority.

OPR/ Name of the operator, if it is not obvious from the aircraft identification in Item 7.

STS/ Reason for special handling by ATS

e.g. Hospital aircraft STS/HOSP
One engine inoperative STS/ONE ENG INOP

TYP/	The type(s) of aircraft, preceded if necessary by numbers of aircraft, if ZZZZ is inserted in Item 9.
PER/	Aircraft performance data, if so prescribed by the appropriate ATS authority
COM/	Significant data related to communication equipment as required by the appropriate ATS authority. e.g. COM/UHF ONLY
DAT/	Significant data related to data link capability using one or more of the letters S, H, V, and M. e.g. DAT/S for satellite data link DAT/H for HF data link DAT/V for VHF data link DAT/M for SSR Mode S data link
NAV/	Significant data related to navigation equipment as required by the appropriate ATS authority. e.g. NAV/INS
DEP/	Name of the departure aerodrome, if ZZZZ is inserted in Item 13, or the ICAO four letter location indicator of the ATS location from which the supplementary flight plan data can be obtained, if AFIL is inserted in Item 13.
DEST/	Name of the destination aerodrome if ZZZZ is inserted in Item 16.
ALTN/	Name of the alternate aerodrome(s) if ZZZZ is inserted in Item 16.
RALT/	Name of enroute alternate aerodrome(s).
RMK/	Any other plain language remarks when required by the appropriate ATS authority or deemed necessary.
DOF/	If a flight plan for a flight conducted wholly in the EUR region is filed more than 24 hours in advance of the EOBT, it is mandatory to provide the date of flight. If the flight plan is less than 24 hours in advance of the EOBT, the date of flight may be optionally indicated. The date is inserted in a six-figure format after the oblique stroke following the DOF indicator: e.g. DOF/YYMMDD

ITEM 19 – SUPPLEMENTARY INFORMATION

This information is not normally included in the transmission of the flight plan. It is retained at the location of the filing of the flight plan in case it is needed.

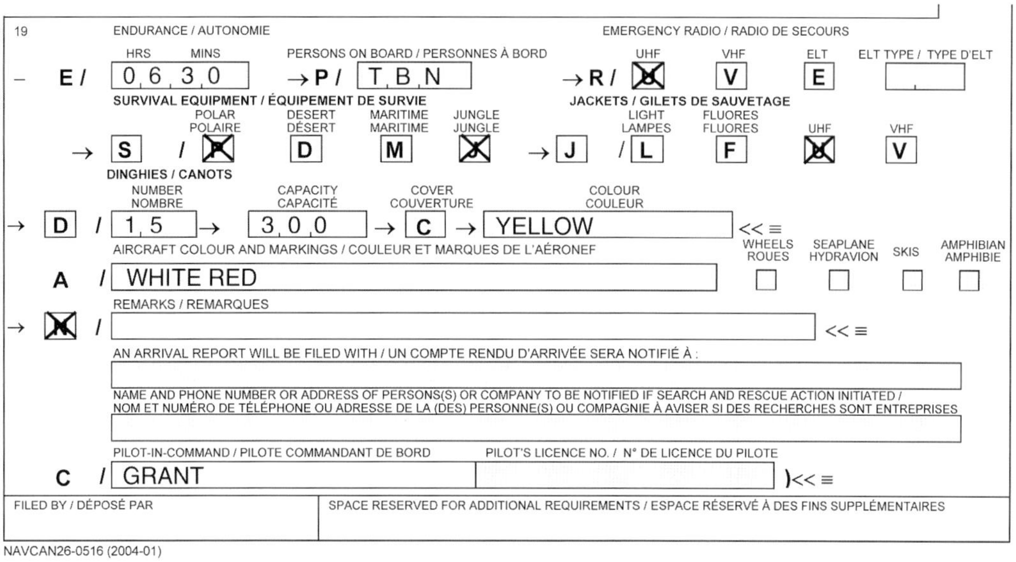

19 ENDURANCE / AUTONOMIE — EMERGENCY RADIO / RADIO DE SECOURS

HRS MINS — PERSONS ON BOARD / PERSONNES À BORD — UHF VHF ELT — ELT TYPE / TYPE D'ELT

– E / 0 6 3 0 → P / T B N → R / U V E

SURVIVAL EQUIPMENT / ÉQUIPEMENT DE SURVIE: POLAR / POLAIRE, DESERT / DÉSERT, MARITIME / MARITIME, JUNGLE / JUNGLE — JACKETS / GILETS DE SAUVETAGE: LIGHT / LAMPES, FLUORES / FLUORES, UHF, VHF

→ S / P D M J → J / L F U V

DINGHIES / CANOTS: NUMBER / NOMBRE, CAPACITY / CAPACITÉ, COVER / COUVERTURE, COLOUR / COULEUR

→ D / 1 5 → 3 0 0 → C → YELLOW << ≡

AIRCRAFT COLOUR AND MARKINGS / COULEUR ET MARQUES DE L'AÉRONEF — WHEELS / ROUES, SEAPLANE / HYDRAVION, SKIS, AMPHIBIAN / AMPHIBIE

A / WHITE RED

REMARKS / REMARQUES

→ N / << ≡

AN ARRIVAL REPORT WILL BE FILED WITH / UN COMPTE RENDU D'ARRIVÉE SERA NOTIFIÉ À :

NAME AND PHONE NUMBER OR ADDRESS OF PERSONS(S) OR COMPANY TO BE NOTIFIED IF SEARCH AND RESCUE ACTION INITIATED / NOM ET NUMÉRO DE TÉLÉPHONE OU ADRESSE DE LA (DES) PERSONNE(S) OU COMPAGNIE À AVISER SI DES RECHERCHES SONT ENTREPRISES

PILOT-IN-COMMAND / PILOTE COMMANDANT DE BORD — PILOT'S LICENCE NO. / N° DE LICENCE DU PILOTE

C / GRANT)<< ≡

FILED BY / DÉPOSÉ PAR — SPACE RESERVED FOR ADDITIONAL REQUIREMENTS / ESPACE RÉSERVÉ À DES FINS SUPPLÉMENTAIRES

NAVCAN26-0516 (2004-01)

Endurance — After E/ insert a 4 figure group giving the fuel endurance in hours and minutes

Persons on Board — After P/ insert the total number of persons (passengers and crew) on board, when required by the appropriate ATS authority,

Insert TBN if the total number of persons is not known at the time of filing.

Emergency and Survival Equipment

R/(Radio) Cross out U if UHF frequency 243.00 MHz is not available
Cross out V if VHF frequency 121.500 MHz is not available
Cross out E if emergency location transmission (ELT) is not available

S/(Survival Equipment)
Cross out all indicators if survival equipment is not carried
Cross out P if polar survival equipment is not carried
Cross out D if desert survival equipment is not carried
Cross out M if maritime survival equipment is not carried
Cross out J if jungle survival equipment is not carried

J/(Jackets) Cross out all indicators if life jackets are not carried
Cross out L if life jackets are not equipped with lights
Cross out F if life jackets are not equipped with fluorescein
Cross out U or V or both as in R/ above to indicate the radio capability of jackets if any

D/(Dinghies)	
Number	Cross out indicators D and C if no dinghies are carried, or insert number of dinghies carried.
Capacity	Insert the total capacity, in persons, of all dinghies carried.
Cover	Cross out indicator C if dinghies are not covered.
Colour	Insert colour of dinghies if carried.
A/(Aircraft Colour and Markings)	Insert the colour of aircraft and significant markings.
N/(Remarks)	Cross out indicator N if no remarks, or indicate any other survival equipment carried and any other remarks regarding survival equipment.
C/(Pilot)	Insert the name of the PIC.

Chapter 11
Jeppesen Airway Manual - Terminal

INTRODUCTION

The terminal section consists of individual aerodrome procedures such as:

- The area chart
- Standard Terminal Arrival (STAR) charts
- Standard Instrument Departure (SID)
- Approach charts
- Supplementary charts that show subjects such as:
  - Noise abatement procedures
  - Airport and parking charts
  - Taxi routings
  - Docking procedures
  - JAA Minimums
  - VFR flight procedures

To illustrate each of these subject areas, this chapter uses Amsterdam, Schiphol. Each chart contains a reference number in a box. The area chart, which is the first chart, has the reference number 10-1 as illustrated:

10-1

AREA CHART (10-1)

The area chart uses exactly the same symbols as the enroute charts. The symbols shown on the following page apply to the area chart:

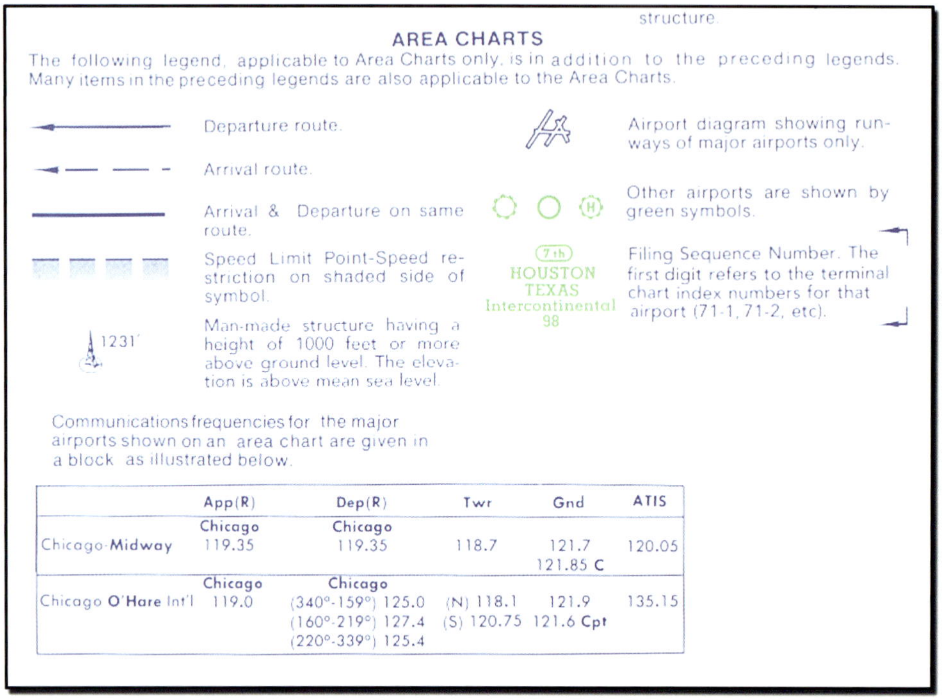

structure.

AREA CHARTS

The following legend, applicable to Area Charts only, is in addition to the preceding legends. Many items in the preceding legends are also applicable to the Area Charts.

Departure route.

Arrival route.

Arrival & Departure on same route.

Speed Limit Point-Speed restriction on shaded side of symbol.

1231′ Man-made structure having a height of 1000 feet or more above ground level. The elevation is above mean sea level.

Airport diagram showing runways of major airports only.

Other airports are shown by green symbols.

7th HOUSTON TEXAS Intercontinental 98 — Filing Sequence Number. The first digit refers to the terminal chart index numbers for that airport (71-1, 71-2, etc).

Communications frequencies for the major airports shown on an area chart are given in a block as illustrated below.

	App(R)	Dep(R)	Twr	Gnd	ATIS
Chicago-Midway	Chicago 119.35	Chicago 119.35	118.7	121.7 121.85 C	120.05
Chicago O'Hare Int'l	Chicago 119.0	Chicago (340°-159°) 125.0 (160°-219°) 127.4 (220°-339°) 125.4	(N) 118.1 (S) 120.75	121.9 121.6 Cpt	135.15

Where terrain is more than 4000 ft above the main airport, terrain information may be depicted. Where terrain information is not shown, this does not mean it is irrelevant. Always refer to the minimum altitudes given for the route structure.

Terminal Exercise 1 — Use chart 10-1

Question 1 Using the communications block for **Schiphol**. After the frequency 126.67 is an X, what does this mean?

Question 2 What is the upper level of Schiphol TMA?

Question 3 What is the elevation of Schiphol airport?

Question 4 When holding at SPY (N52 32.4 E004 51.2) what is the inbound track?

Question 5 Can an aircraft follow the route BERGI B5 SPY?

STANDARD TERMINAL ARRIVAL (STAR)

Using chart 10-2, note the top corner contains a box telling you what the chart is. For example:

STAR

A SID/STAR legend is on pages 81 to 84 of the Introduction. A circle with an approximate runway plan highlights the terminal airport SCHIPHOL. The ATIS frequencies, 108.4 MHz and 132.975 MHz, appear at the top of the chart.

The Transition Level and Transition Altitude values are in the top left-hand corner of the chart.

Within the box in the centre top are the arrivals that apply:

EELDE A, REKKEN A

EELDE B, REKKEN B
BY ATC

ARRIVALS
(RWYS 01R, 06, 19R, 27)

The names of the arrivals usually refer to a radio aid, in this case EEL and REK.

The arrival routes are in a plan format. Text boxes on the chart provide any required explanatory material.

Terminal Exercise 2 Use chart 10-2.

Question 1 What is the distance between EEL and ARTIP?

Question 2 What is the holding speed at SPY?

Question 3 What are the entry levels into the Schiphol TMA?

Question 4 What is the maximum speed at SPL 15 nm?

Question 5 What is the IAS for an aircraft approaching RKN from REMKO?

STANDARD INSTRUMENT DEPARTURE (SID)

Use chart 10-3. This chart uses the same legend and format as the STAR.

Terminal Exercise 3 Use chart 10-3.

Question 1 What angle of bank is expected in a turn?

Question 2 At what altitude should an aircraft contact Schiphol Departure?

Question 3 For a 22 departure, after PAM what routing must aircraft follow initially?

Question 4 For a departure on 19L, what is the initial track?

Question 5 What is the maximum IAS for a 22 departure?

Question 6 For a 06 departure at what radial does an aircraft turn to intercept the 273° radial PAM?

Question 7 If an aeroplane is to proceed along UR12:

a. What altitude should the aircraft be at or above at D60 DHE?
b. What is the frequency of DHE?

APPROACH CHART

Use chart 11-1, ILS Rwy 06, NDB DME Rwy 06. The approach chart follows a standard format.

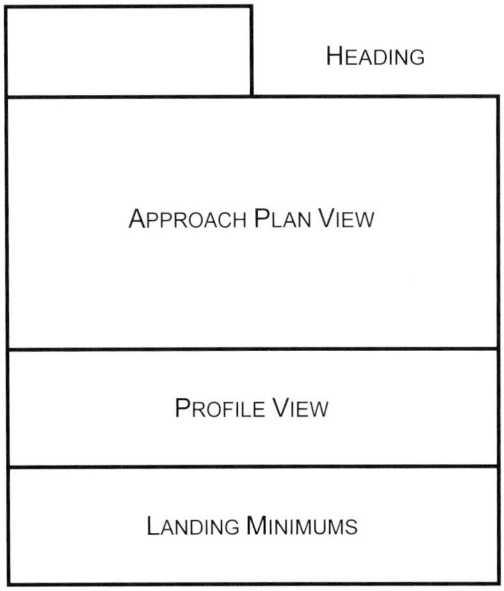

Approach chart legends are on pages 102 to 115 of the Introduction. To the right of the top panel are:

➢ The geographical location	**AMSTERDAM, NETHERLANDS**
➢ The airport name	**SCHIPHOL**
➢ The procedure identification	**ILS Rwy 06**
	NDB DME Rwy 06
➢ The primary facility frequency and identifier	**LOC 110.55 SL**
➢ Airport elevation	**-11 ft (below sea level)**

An MSA circle is centred upon the major aid near the airfield, SPL VOR/DME. Heights and limits are shown.

The communications and altimeter setting data are on the right side of the top panel.

The plan view is a graphic picture of the approach at a scale of 1 inch = 5 nm. Other scales are illustrated on the chart.

The plan of the approach takes over where the STAR stops.

The profile view gives a vertical profile of the expected approach.

The minima section gives the minimum DA(H) for each specified approach. Categories of aircraft may have different RVR/Visibility limits, listed below the DA(H).

Terminal Exercise 4 Use chart 11-1

Question 1 At what DME is the MAP from SPL?

Question 2 In the communications listing for Schiphol Approach, what does the (R) mean?

Question 3 On a NDB DME approach at 6 DME what height should the aircraft be passing on the QFE set?

Question 4 What is the minimum visibility for a circling approach at 180 knots?

Question 5 What is the OCH for a category B aircraft?

Question 6 At what distance is the outer marker from the threshold?

Question 7 What does ALS mean when looking at the minima section?

Question 8 What is the altitude that you would expect to be at the LCTR outbound?

Question 9 How long is the outbound leg from the LCTR?

Question 10 At what DME from SPL would you start the descent?

Question 11 What does TCH 55 ft mean?

SUPPLEMENTARY PAGES

Noise Abatement Procedures (Charts 10-4 to 10-4B) The noise abatement procedures are designed to avoid excessive aircraft noise in the vicinity of an airfield. The Amsterdam procedures follow a simple format which includes the following:

- Runway usage
- Preferential runway system
- Arrivals
- Departures
- Night time restrictions
- Reverse thrust APUs

Terminal Exercise 5 Use chart 10-4 to 10-4B

Question 1 What is the maximum tailwind component for a dry runway?

Question 2 What is the minimum climb gradient for runway 01 and to what altitude?

Question 3 What is the preferred sequence for selecting runways for take-off?

Question 4 When runways 19R and 27 are being used for arrivals, what is the minimum visibility and cloud base?

Question 5 What are the initial procedures for a jet aircraft on take-off to 1500 ft?

Question 6 When can reverse thrust be used on Rwy 06?

Airport Charts

The manual provides various airport charts.

Chart 10-8 is on yellow paper. This means that the entry is temporary. Some charts may refer the reader to active NOTAMs.

Normally some explanation is given. In this case, it is due to work on the DE apron.

Chart 10-9 shows the airport plan major roads and rivers. While both are grey on the chart, they can easily be discerned. The runway plan shows the parking areas. Restrictions are listed on the chart, such as the maximum wingspan for entry to the apron via the EAST taxiway – 171 ft.

The airport legend has a full decode, which can be found on pages 116 to 119 of the introduction.

Chart-9A lists the dependent and independent landing runways with their relevant restrictions as well as general landing information.

Chart-9A1 covers the low visibility procedures, start up procedure, and push-back and taxi procedures.

Chart-9A2 lists ways to reduce runway occupancy times for the different classes of aircraft.

Chart 10-9B/C are expanded charts that show the taxi routes for both after landing and before take-off with the relevant controlling frequency.

Chart 10-9D/E is a list of all stands and their respective INS co-ordinates.

Chart 10-9F/G lists the visual docking guidance system and the indications.

Chart 10-9X/X1 and X2 lists the minima for all runways.

Examine and study the airport charts with the legend. Different airports use the same procedures but they may be listed in a slightly different way.

TERMINAL EXERCISE ANSWERS

Terminal Exercise 1

Question 1 On request

Question 2 FL 95

Question 3 -11 ft

Question 4 242°M

Question 5 No, as the airway is normally one way westbound

Terminal Exercise 2

Question 1 56 nm

Question 2 250 knots IAS

Question 3 A maximum at SPE 30 nm of FL 100, with a minimum level of FL 70 at the TMA boundary unless otherwise instructed

Question 4 220 knots

Question 5 280 to 300 knots

Terminal Exercise 3

Question 1 25°

Question 2 When passing 2000 ft

Question 3 017R to ANDIK, intercept the 054R SPY to GRONY

Question 4 185°M

Question 5 220 knots

Question 6 103°M

Question 7

a. FL 260
b. 116.3 MHz

Terminal Exercise 4

Question 1 2.6 nm

Question 2 Radar available

Question 3 1292 ft

Question 4 3600 m

Question 5 154 ft

Question 6 3.9 nm

Question 7 Approach lights out of service

Question 8 3000 ft

Question 9 1 minute

Question 10 8.2 nm

Question 11 The threshold crossing height is 55 feet

Terminal Exercise 5

Question 1 5 knots

Question 2 5%/150 ft

Question 3 24, 01L, 19L, 09

Question 4 3000 m visibility, cloud base 1000 ft or more

Question 5 Take-off power, Take-off flaps. Climb at V_2 +10 knots (as limited by pitch angle)

Question 6 0700 – 2300 LT

Chapter 12

Jeppesen Airway Manual - Jeppesen VFR + GPS Chart, Germany ED-6

INTRODUCTION

This chart is used in the flight planning examination for simple VFR plotting and extraction of data.

The chart is designed for planning and conducting flights in VMC conditions and in accordance with VFR rules.

CHART INFORMATION

The front panel shows the coverage of the chart and the adjacent charts. Note that the ED-6 is effective below:

- FL 125 in Austria
- FL 115 in France
- FL 100 in Germany
- FL 150 in Switzerland

GPS LATITUDE AND LONGITUDE DISCREPANCIES

GPS position is based upon WGS-84. Some governments still base position information upon local geodetic reference datums. On the chart, all positions listed on the right hand panels are referenced to WGS-84:

- VFR reporting points
- Aerodromes
- Radio Navigation Aids

AERONAUTICAL INFORMATION

Within the aeronautical information panels are the chart symbols and explanations.

Along the bottom of the chart are further explanations of:

- Flight Information and meteorological services
- General Aviation Forecast Areas (the numbers refer to the station telephone numbers)
- Airspace classification – Germany
- Airspace classification in Germany for VFR

- Phonetic alphabet and Morse code
- Feet/Metre conversion
- Airspace designators and control frequencies

Note that the scale of the map is calibrated in kilometres, nautical miles and statute miles. Elevations and Minimum Grid Area Altitudes are given in feet.

PROJECTION

Down the left-hand side of the chart the projection and standard parallels are listed.

- Lamberts Conformal
- Standard parallels of 37°N and 65°N

Complete the following questions with reference to the ED-6

VFR Question 1 Within the Munich CTR (Munich N48 21.2 E011 47.2) is the point ***FOXTROTT 2***. What is Foxtrott 2 and what is its position?

VFR Question 2 What is the ATIS frequency for Augsburg?

VFR Question 3 What type of airport is Tannheim (N48 00.7 E010.06.1)?

VFR Question 4 Decode the navaids at positions:

a. N49 08.6 E010 14.3
b. N48 21.9 E007 49.7
c. N49 13.7 E007 25.0

VFR Question 5 What is the classification of the Neuberg AB Airspace (N48 42.7 E011 12.7)? What is the upper limit of this airspace?

VFR Question 6 Decode the symbols at:

a. N47 14 E009 42
b. N47 49.5 E00731
c. N48 40 E009 07
d. N47 56 E013 25.8

VFR Question 7 What is the bearing and distance of NORDLINGEN (N48 52.4 E010 30.3) from GERSTETTEN (N48 37.3 E010 03.7)?

VFR Question 8 What VFR classes of airspace are in use in Germany?

VFR Question 9 For a VFR route in France, how are distance and bearing measured?

VFR Question 10 For Zurich (N47 27.5 E008 32.9), what is:

a. the ICAO four letter identifier?
b. the elevation?
c. the runway available?

VFR Question 11	Next to the frequency of 118.10 at Zurich is (v), what does this mean?
VFR Question 12	What is the magnetic variation at Zurich?
VFR Question 13	To what year are the isogonals accurate?
VFR Question 14	What does GAFOR mean?
VFR Question 15	What is the line that crosses N49 00 E010 17.5 in a NE-SW direction?
VFR Question 16	What is the distance in kilometres between Stuttgart (N48 42.7 E009 20.1) and Luburg (N48 54.8 E009 20.4)?
VFR Question 17	What is the Munich Flight Information frequency?
VFR Question 18	What is the frequency of Hochwald VOR/DME?
VFR Question 19	To what altitude is Class E airspace in the Alps region of Germany?
VFR Question 20	At Laupheim what radio navigation aids are available and on what frequency?

VFR ANSWERS

VFR Question 1 VFR Reporting Point
N48 27.5 E011 48.6 or 248°/19 MBG

VFR Question 2 124.575 MHz

VFR Question 3 Civil airport

VFR Question 4

a. N49 08.6 E010 14.3 Dinkelsbuhl VORTAC 117.80 MHz DKB
b. N48 21.9 E007 49.7 Lahr DME 108.05 MHz LRD
c. N49 13.7 E007 25.0 Zweibrucken NDB 435 KHz ZBN

VFR Question 5 Class D, 3700 ft. (Note that this is classed as a part time CTR – not open 24 hours a day)

VFR Question 6 Decode the symbols at:

a. Hang glider
b. VFR and TMA transit route with waypoint
c. NDB
d. Group of obstructions

VFR Question 7 049°T/23.5 nm

VFR Question 8 Classes C, D, E, F, G
VFR Question 9 Nautical miles and °M

VFR Question 10

a. LSZH
b. 1416 ft
c. 1000 m

VFR Question 11 VDF available

VFR Question 12 0°

VFR Question 13 1999

VFR Question 14 General Aviation Forecast Areas

VFR Question 15 FIR boundary

VFR Question 16 23 kilometres

VFR Question 17 126.95 MHz

VFR Question 18 113.20 MHz HOC

VFR Question 19 FL 130

VFR Question 20 NDB 407 LUP

Chapter 13
Meteorological Messages

INTRODUCTION

A variety of weather messages are originated by Meteorological Observers at aerodromes. These are collated and broadcast in text form to stations around the world. Aviators should be able to distinguish between the various message types and their uses.

AERODROME METEOROLOGICAL REPORT

Aerodrome Meteorological Reports (METAR) contain observations on the conditions that actually exist at a station and are made every 30 minutes throughout the day.

a. Short term landing forecasts, valid for two hours (TREND), may be added to METARS.
b. Information on runway condition is added to METAR when appropriate, until these conditions have ceased.

SPECIAL AERODROME METEOROLOGICAL REPORTS

Special Aerodrome Meteorological Reports (SPECI) are issued when conditions change significantly. Selected Special Reports (SPECI) are defined as Special Reports disseminated beyond the aerodrome of origin.

TERMINAL AERODROME FORECASTS

Terminal Aerodrome Forecasts (TAF) are normally provided only for those aerodromes where official meteorological observations are made. Local Area Forecasts are provided for other aerodromes. Amended TAFs or Local Area Forecasts are issued when forecast conditions change significantly.

ACTUAL WEATHER CODES

The content and format of an actual weather report is shown in the following table.

Report Type	Location Identifier	Date/Time	Wind	Visibility	RVR
METAR	EGSS	291250Z	31015G30KT	1400SW 6000N	R24/P1500

Present Weather	Cloud	Temp/ Dew Pt	QNH	Recent Weather	Wind Shear	Trend	Rwy State
SHRA	FEW005 SCT010CB BKN025	10/05	Q0999	RETS	WS RWY25	NOSIG	88290592

IDENTIFIER

The identifier has three components:

Report Type: METAR or SPECI.
ICAO Indicator: This is a four-letter group indicating the airfield, eg, EGPL, LFPB.
Date/Time UTC: In a METAR or SPECI, this is the date and time of the observation in hours and minutes UTC, eg 091250Z.
Example: METAR EGDL 211020Z.

Note: If a meteorological bulletin consists of a set of reports from one or more airfields, the codename METAR or SPECI may be replaced by:

- SA (Actual Report), or
- SP (Special Report)

followed by a bulletin identifier, date and time of the observation.

SURFACE WIND VELOCITY

The first three figures indicate the wind direction (T) to the nearest 10°, followed by two figures (exceptionally three figures) giving the mean windspeed during the previous ten minutes. The permitted units of speed are:

- KT indicating knots
- KMH for kilometres per hour, or
- MPS for metres per second.

Example: 30015KT

These may be followed by a letter G and two more figures if the maximum gust speed exceeds the average speed by 10kt or more.

Example: 30015G30KT.

Variations in wind direction of 60° or more in the ten minutes preceding the observation are shown as three figures then the letter V followed by another three figures but only if the speed is more than 3kt

Example: 270V330 meaning, the wind is varying in direction between 270°T and 330°T.

00000 indicates calm conditions and a variable wind direction is shown by VRB followed by the speed.

HORIZONTAL VISIBILITY

When there is no marked variation in direction, the minimum visibility is given in metres. The minimum visibility with the direction is given when there is a marked variation with direction,

Example: 2000NE.

When the minimum visibility is less than 1500 metres and the visibility in any other direction is greater than 5000 metres, the maximum visibility and its direction is also shown.

Example: 1200NE 6000SW.

9999 indicates a visibility of 10 kilometres or more, 0000 indicates a visibility of less than 50 metres.

RUNWAY VISUAL RANGE (RVR)

Runway Visual Range is reported when the meteorological visibility falls below 1500 m. It has the form R, followed by the runway designator, a diagonal and then the Touchdown RVR. If more than one runway is in use, the RVR group is repeated. Parallel runways are distinguished by adding C, L, or R to the runway designator.

Example: R24L/1200R24R/1100.

When RVR is greater than the maximum assessable value the prefix P will be added followed by the maximum value.

Example: R15/P1500.

The prefix M indicates the RVR is less than the minimum value that can be assessed.

Example: R15/M0050.

Tendencies are indicated by U for up, D for down or N for no change. They show a significant change (100 m or more) from the first five minutes to the second five minutes in the ten minute period prior to the observation.

Example: R25/1000D.

Variations are reported if the RVR has changed minute by minute during the ten minute period prior to the report. The one minute minimum and maximum separated by V are reported instead of the ten minute mean.

Example: R15L/0850V1000.

WEATHER

Each weather group may consist of the appropriate intensity indicators and abbreviations; making groups of two to nine characters from the table below.

SIGNIFICANT PRESENT AND FORECAST WEATHER CODES

QUALIFIER		WEATHER PHENOMENA		
Intensity or Proximity	**Descriptor**	**Precipitation**	**Obscuration**	**Other**
1	2	3	4	5
- Light	MI Shallow	DZ Drizzle	BR Mist	PO dust/sand whirls
Moderate (no qualifier)	BC Patches	RA Rain	FG Fog	
	PR Partial (Covering part of Aerodrome)	SN Snow	FU Smoke	SQ Squalls
+ Heavy 'Well developed in the case of FC and PO'	DR Drifting	SG Snow Grains	VA Volcanic Ash	FC Funnel Cloud(s) (tornado or water-spout)
VC In the vicinity (within 8km of aerodrome perimeter but not at aerodrome)	BL Blowing	IC Ice Crystals (Diamond Dust)	DU Widespread Dust	SS Sandstorm
	SH Shower(s)	PE Ice-Pellets	SA Sand	DS Duststorm
	TS Thunderstorm	GR Hail	HZ Haze	
	FZ Freezing Super-Cooled	GS Small hail (<5 mm diameter) and/or snow pellets		

A mixture of weather can be reported using up to three groups to indicate different weather types.

Examples: MIFG, VCBLSN, +SHRA, -DZHZ

Note: BR, HZ, FU, IC, DU and SA will not be given in METAR or TAF when the visibility is above 5000 m.

CLOUD

The cloud group usually consists of three letters and three figures. These show the cloud amount followed by the height of the cloudbase, above airfield level, in hundreds of feet. The cloud groups are given in ascending order of height.

Example: SCT015 or OVC080.

These groups are:

FEW indicating 1-2 oktas
SCT (scattered) indicating 3-4 oktas
BKN (broken) indicating 5-7 oktas
OVC (overcast) indicating 8 oktas.

The cloud group may have a suffix for significant convective cloud, CB for Cumulonimbus or TCU for Towering Cumulus. No other cloud types are reported.

Example: BKN015CB.

Layers are reported as:

First Group	Lowest individual layer of any amount
Second Group	Next individual layer of more than 2 oktas
Third Group	Next higher layer of more than 4 oktas
Additional Group	Significant convective cloud not already reported.

SKC indicates no cloud to report when CAVOK does not apply.

Sky obscured is shown by VV followed by vertical visibility in hundreds of feet. When the vertical visibility cannot be assessed the group will read VV///.

Example: VV003

CAVOK

CAVOK is used in place of groups 4, 5, 6, and 7 when all of the following conditions apply:

a. Visibility is 10 km or more.
b. There is no cloud below 5000 ft or below the highest Minimum Sector Altitude (MSA), which ever is greater, and no CB.
c. No significant weather phenomenon at or in the vicinity of the aerodrome.

Minimum Sector Altitude is the lowest altitude that may be used under emergency conditions. It provides a minimum clearance of 1000 ft above all objects located in an area contained within a sector of a circle of 25 nm radius centred on a radio navigation aid. A sector cannot be less than 45°.

AIR TEMPERATURE AND DEWPOINT

Air Temperature and Dewpoint are reported in degrees Celsius. M indicates a negative value.

Examples: 10/08, 01/M01

SEA LEVEL PRESSURE (QNH)

QNH is reported in the form Q followed by a four figure group. If the QNH is less than 1000 mb the first figure is a 0. QNH is rounded down to the nearest whole millibar.

Example: Q0995

The pressure may be given in inches of mercury. Then it is reported as A followed by the pressure in hundredths of inches.

Example: A3037

SUPPLEMENTARY INFORMATION

RECENT WEATHER (RE)

This is operationally significant weather observed since the previous observation (or in the last hour, whichever is the shorter) but not occurring now. Up to three groups may be used to indicate the former presence of more than one weather type.

Example: RETS REGR

WINDSHEAR (WS)

If reported, Windshear may be inserted in the lowest 1600 feet of the take-off or approach paths.

Example: WS RWY27, WS ALL RWY

TREND

A forecast of significant changes in weather expected within two hours of the observation time may be added to the end of a METAR or SPECI if a qualified Forecaster is present.

Change Indicator:

BECMG (becoming) or TEMPO (temporary) which are followed by a time group in hours and minutes UTC, and which may be followed by FM (from), TL (until) or AT (at) followed by a four figure time group.

Weather:

Standard codes are used in this section. NOSIG is used when no significant changes are expected to occur during the trend forecast period.

Example: BCMG FM1100 25035G50KT or, TEMPO 0630 TL 0830 3000 SHRA.

Only those elements of the above in which a change is expected are included. When no change is expected, the term NOSIG is used.

RUNWAY STATE GROUP

An eight figure Runway State Group may be added to the end of the METAR or SPECI (following any TREND) when there is lying precipitation or other runway contamination. The first two digits are the runway designator, and last two digits are braking action. The complete group consists of:

Runway Designator (First Two Digits)

27 = Runway 27 or 27L

77 = Runway 27R (50 added to the designator to indicate 'right' Runway)

88 = All runways

99 = A repeat of last message because no new information received

Runway Deposits (Third Digit)

0 = Clear and dry
1 = Damp
2 = Wet or water patches
3 = Rime or frost covered
(depth normally less than 1 mm)
4 = Dry Snow
5 = Wet Snow
6 = Slush
7 = Ice
8 = Compacted or rolled snow
9 = Frozen ruts or ridges
/ = Not reported (eg due to runway clearance in progress)

Extent of Runway Contamination (Fourth Digit)

1 = 10% or less
2 = 11% to 25%
5 = 26% to 50%
9 = 51% to 100%
/ = Not reported (eg due to runway clearance in progress)

Depth of Deposit (Fifth and Sixth Digits) The quoted depth is the mean number of readings or, if operationally significant, the greatest depth measured.

00 = less than 1 mm	01 = 1 mm through to 90 = 90 mm
91 = not used	92 = 10 cm
93 = 15 cm	94 = 20 cm
95 = 25 cm	96 = 30 cm
97 = 35 cm	98 = 40 cm or more

// = Depth of deposit operationally not significant or not measurable

Friction Coefficient or Braking Action (Seventh and Eighth Digits) The value transmitted is the mean or, if operationally significant, the lowest value.

28 = Friction coefficient 0.28	35 = Friction coefficient 0.35
or	
91 = Braking action: Poor	92 = Braking action: Medium/Poor
93 = Braking action: Medium	94 = Braking Action: Medium/Good
95 = Braking action: Good	

99 = Figures unreliable (e.g. if the equipment used does not measure satisfactorily in slush or loose snow)

// = Braking action not reported (e.g. runway not operational; closed; etc)

If contamination conditions cease to exist, the abbreviation CLRD is used.

Examples: 24CLRD93 = Rwy 24 cleared: Braking action; Medium
88CLRD95 = All runways cleared: Braking action; Good

'AUTO' AND 'RMK'

Where a report contains fully automated observations with no human intervention, it is indicated by the code word 'AUTO,' inserted immediately before the wind group.

The indicator 'RMK' (remarks) denotes an optional section containing additional meteorological elements. It is appended to METARs by national decision, and is not disseminated internationally.

MISSING INFORMATION

Information that is missing from a METAR or SPECI may be replaced by diagonals.

EXAMPLES OF METARS:

SAUK02 EGGY 301220Z METAR

EGLY 24015KT 200V280 8000 -RA SCT010 BKN025 OVC080 18/15 Q0983 TEMPO 3000 RA BKN008 OVC020=

EGPZ 30025G37KT 270V360 1200NE 6000S +SHSN SCT005 BKN010CB 03/M01 Q0999 RETS WS LDG RWY27 BECMG AT 1300 9999 NSW SCT015 BKN100=

The METARs above are for 1220 UTC on the 30th day of the month. The decode in plain language is:

EGLY: Surface wind: mean 240°True, 15 kt; varying between 200° and 280° minimum visibility 8 km; slight rain; cloud: 3-4 oktas base 1000 ft, 5-7 oktas 2500 ft, 8 oktas 8000 ft; Temperature +18°C, Dew Point +15°C; QNH 983 mb; Trend: temporarily 3000 m in moderate rain with 5-7 oktas 800 ft, 8 oktas 2000 ft.

EGPZ: Surface wind: mean 300°True, 25 kt; maximum 37 kt, varying between 270° and 360°; minimum vis 1200 m (to northeast), maximum visibility 6 km (to south); heavy showers of snow, Cloud: 3-4 oktas base 500 ft, 5-7 oktas CB base 1000 ft; Temperature +3°C, Dew Point -1°C; QNH 999 mb; thunderstorm since previous report; windshear reported on approach to runway 27; Trend: improving at 1300 UTC to 10 km or more, nil weather, 3-4 oktas 1500 ft, 5-7 oktas 10 000 ft.

AERODROME FORECASTS (TAF) CODES

TAF describes the forecast of conditions at aerodromes and usually cover periods of not less than 9 hours, and not more than 24 hours. Those valid for less than 12 hours are issued every 3 hours and those valid for 12 to 24 hours are issued every 6 hours. TAFs prefixed FC are valid for periods of less than 12 hrs. TAF's prefixed FT are valid for periods of 12 to 24 hours. An 18 hour forecast normally starts 8 hours after the time of issue and normally accompanies a 9 hour TAF.

TAF CONTENTS AND FORMAT

The TAF uses the same code system as the METAR, with the following differences:

Validity Period — In the validity period the first two numbers indicate date of issue. The next four figures the forecast period in whole hours UTC. If the TAF bulletin consists of forecasts for one or more airfields, the codename TAF may be replaced by FC or FT, followed by the date and time of origin and neither codename nor time/date group will appear in the forecast.

Visibility — Same as METAR with only the minimum visibility forecast.

Weather — If no significant weather is expected, the group is omitted. After a change group, if the weather becomes insignificant, NSW (No Significant Weather) is used.

Cloud — If clear sky is forecast, the cloud group is replaced by SKC (Sky Clear). If CAVOK and SKC are not appropriate, NSC (No Significant Cloud) is used.

SIGNIFICANT CHANGES

FM followed by the time to the nearest hour and minute UTC, is used to show the beginning of a self contained part in the forecast. All conditions given before this group are superseded - they no longer apply.

Example: FM1220 27017KT 4000 BKN010.

BCMG followed by a four figure time group indicating the earliest and latest start hours of an expected permanent alteration to the meteorological conditions. This change can occur at a regular or irregular rate during the forecast change period. The change will not start before the first time and is complete by the second time given.

Example: BECMG 2124 1500 BR.

TEMPO followed by a four figure time group indicates the hours of a period of changes in the conditions of a temporary nature which may occur at any time during the period. These changes are expected to last less than one hour in each case and in total for less than half of the forecast period indicated.

PROBABILITY of the occurrence of alternative forecast conditions is given as a percentage but only 30% or 40% is used.

Example: PROB30 0507 0800 FG BKN004
PROB40 TEMPO 1416 TSRA BKN010CB.

OTHER GROUPS

Three additional TAF groups may be used in overseas and UK military TAF. They are used to forecast temperature (Group indicator T), Icing (Group indicator 6) and turbulence (Group indicator 5).

EXAMPLE 9 HR TAF

FCUK33 EGGY 300900Z

EGGW 301019 23010KT 9999 SCT010 BKN018 BECMG 1114 6000 -RA BKN012 TEMPO 1418 2000 RADZ OVC004 FM1800 30020G30KT 9999 -SHRA BKN015CB=

Decode:

Nine hour TAF issued at 0900 UTC on the 30th of the month at Luton, Valid from 1000 to 1900 UTC. Wind from 230°T at a speed of 10 Kt. Visibility 10 kilometres or more. Cloud amount 3 - 4 oktas, base 1000 ft, second cloud layer 5 - 7 oktas, base 1800 ft. Between 1100 - 1400 UTC a permanent change will occur. Visibility will become 6 km in slight rain. with 5 - 7 oktas of cloud base 1200 ft. There will be short term changes between 1400 - 1800 UTC. Visibility will decrease to 2000 metres in moderate rain and drizzle and overcast at 400 ft. From 1800 UTC there will be another permanent change. Wind velocity will become 300°T at 20 kt gusting to 30 kt. Visibility will improve to 10 km or more with slight rain showers and the cloud will be 5 - 7 oktas of cumulonimbus base 1500 ft.

EXAMPLE 18 HR TAF

FTUK31 EGGY 102300Z

EGLL 110624 13010KT 9000 BKN010 BECMG 0608 SCT015 BKN020 PROB30 TEMPO 0816 17025G40KT 4000 TSRA SCT010 BKN015CB BECMG 1821 3000 BR SKC=

Decode:

Eighteen hour TAF issued at 2300 UTC on the 10th for London Heathrow, valid from 0600 - 2400 UTC on the 11th. Wind from 130°T at 10 kt. Visibility 9 km. Cloud 5 - 7 oktas base 1000 ft. A permanent change will occur between 0600 - 0800 UTC to 3 - 4 oktas of cloud base 1500 ft and 5 - 7 oktas of cloud base 2000 ft. There is a 30% probability that for short periods between 0800 - 1600 UTC the wind velocity will become 170°T speed 25 kt maximum to 40 kt with visibility of 4000 m in thunderstorms with rain, cloud becoming 3 - 4 oktas base 1000 ft and 5 - 7 oktas cumulonimbus base 1500 ft. A permanent change will occur between 1800 - 2100 UTC the visibility becoming 3000 m in mist with clear skies.

VOLMET BROADCASTS

These are aerodrome weather reports, METARS, which are transmitted on VHF frequencies in plain language in the following order:

- Surface Wind Velocity (degrees True)
- Visibility
- RVR if applicable
- Weather
- Cloud
- Temperature
- Dewpoint
- QNH
- TREND if applicable, or CAVOK
- The spoken word SNOCLO will be added at the end of the aerodrome report when the aerodrome is unusable for take-offs and landings due to heavy snow on runways or runway clearance operations.

Chapter 14
Upper Air Charts

INTRODUCTION

Middle and upper level charts vary in coverage from FL 100 to FL 630 depending upon the area covered. The layout and symbology used is similar to that taught during the PPL. Other symbology includes:

SYMBOLS FOR SIGNIFICANT WEATHER

Symbol	Meaning	Symbol	Meaning
	Thunderstorms	,	Drizzle
	Tropical cyclone	/// /// /// ///	Rain
	Severe squall line*	*	Snow
	Moderate turbulence	∇	Shower
	Severe turbulence		Widespread blowing snow
	Mountain waves		Severe sand or dust haze
	Moderate aircraft icing		Widespread sandstorm or dust storm
	Severe aircraft icing	∞	Widespread haze
≡	Widespread fog	=	Widespread mist
△	Hail		Widespread smoke
	Volcanic eruption**		Freezing precipitation ***
		■	Visible ash cloud

* In flight documentation for flights operating up to FL 100, this symbol refers to a squall line.

** The following information referring to the symbol should be included in the side of the chart.

- Volcanic eruption
- Name of volcano
- Latitude and longitude
- Date and time of the first eruption
- Check SIGMET for volcanic ash

*** This symbol does not refer to icing due to precipitation coming into contact with an aircraft at a very low temperature.

FRONTS AND CONVERGENCE ZONES AND OTHER SYMBOLS

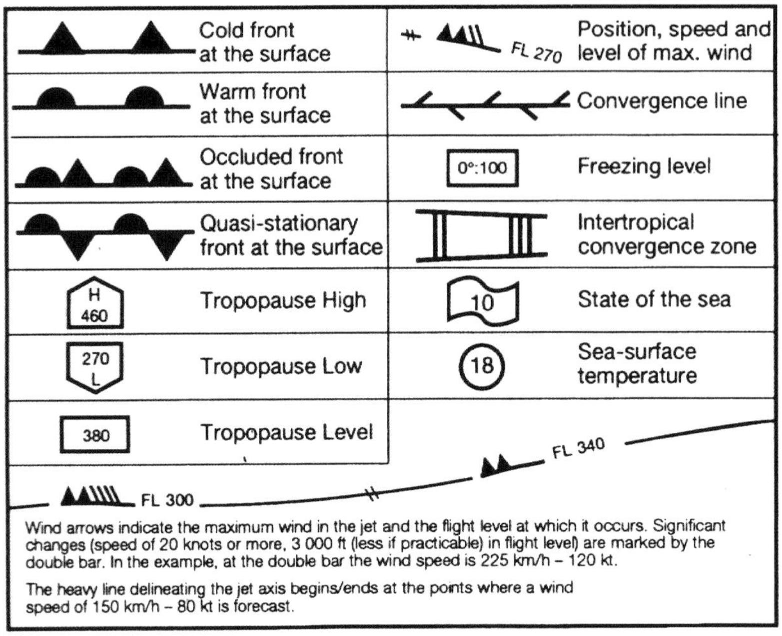

Where the cold front, warm front, occlude front, and quasi-stationary front symbols are not filled in, then the front is above the surface.

The above diagram indicates a cold front above the surface.

CLOUD ABBREVIATIONS

CI	Cirrus	AS	Altostratus	ST	Stratus
CC	Cirrocumulus	NS	Nimbostratus	CU	Cumulus
CS	Cirrostratus	SC	Stratocumulus	CB	Cumulonimbus
AC	Altocumuluus				

CLOUD AMOUNT

Clouds except CB

SKC	Sky clear	$^0/_8$
FEW	Few	$^1/_8$ to $^2/_8$
SCT	Scattered	$^3/_8$ to $^4/_8$
BKN	Broken	$^5/_8$ to $^7/_8$
OVC	Overcast	$^8/_8$

CUMULONIMBUS ONLY

ISOL	Individual CBs (isolated)
OCNL	Well separated CBs (occasional)
FREQ	CBs with little or nor separation (frequent)
EMBD	Thunderstorm clouds contained in layer of other clouds (embedded).

WEATHER ABBREVIATIONS

DZ	Drizzle	GEN	General
LOC	Locally	LYR	Layer
K	Thunderstorm	BLW	Below
COT	At the coast	SEV	Severe
WDSPR	Widespread		
SH	Showers		
FZ	Freezing		
MAR	Over the sea		

LINES AND SYMBOLS ON THE CHART

SPEED OF A FRONT IN KNOTS

SLOW — SPEED OF THE FRONT CAN BE DEPICTED IN WORDS

BOUNDARY OF AREA OF SIGNIFICANT WEATHER

— — BOUNDARY OF AREA OF CLEAR AIR TURBULENCE
THE CAT AREA MAY BE MARKED BY A NUMERAL INSIDE A SQUARE AND A LEGEND DESCRIBING THE NUMBERED CAT AREA MAY BE ENTERED IN A MARGIN

- - - 0° C: FL120 - - - ALTITUDE OF THE 0°C ISOTHERM IN FLIGHT LEVELS

SIGNIFICANT WEATHER CHART

Chart 1 is an example of a high level chart issued by London. The chart covers a considerable area of Europe, the Middle East, North Africa, and North America. These charts are issued in advance of their valid times, which are 0000, 0600, 1200, and 1800 UTC. The validity of this chart being 1200 UTC on 17 August. A Polar Stereographic or Mercator projection is used for all middle and upper air significant weather charts.

NB Take great care when measuring direction on all small scale meteorological charts. Use a square navigation protractor .

a. The bottom right-hand corner of this chart gives a box which:
   i. Indicates the issuing station
   ii. The type of chart – Significant Weather
   iii. The depth of the atmosphere covered. In this case FL 250 – 630. This can be in hPa.
   iv. The chart is a fixed time chart for 1200 UTC, 17 August
   v. The units used on the chart are Pressure Altitude (Hectofeet), knots and °C

b. The bottom right box indicates that all heights are Flight Levels. Tropopause heights are shown in boxes on the chart (Indicated by A on the chart). The symbols and CB imply moderate or severe turbulence and icing.

c. The vertical distance at which phenomena are expected are indicated by flight levels, top over base or top followed by base. 'XXX' means the phenomenon is expected to continue above or below the vertical coverage of the chart (Indicated by B on the chart).

d. The surface positions together with the direction and speed of movement of pressure centres and fronts are denoted as shown on the chart. Where slow is used this indicates movement of less than 5 knots (Indicated by C on the chart).

e. Dashed lines denote areas of CAT. These areas are numbered and associated with the decode box on the chart in the bottom right corner (Indicated by D on the chart).

e.g. Area 4 Moderate turbulence FL 370 to FL 300

f. On lower charts the 0° C Isotherm is also shown as a dotted line with the FL indicated.
e.g. - - - - - - - 0°C:FL130 - - - - - - - -

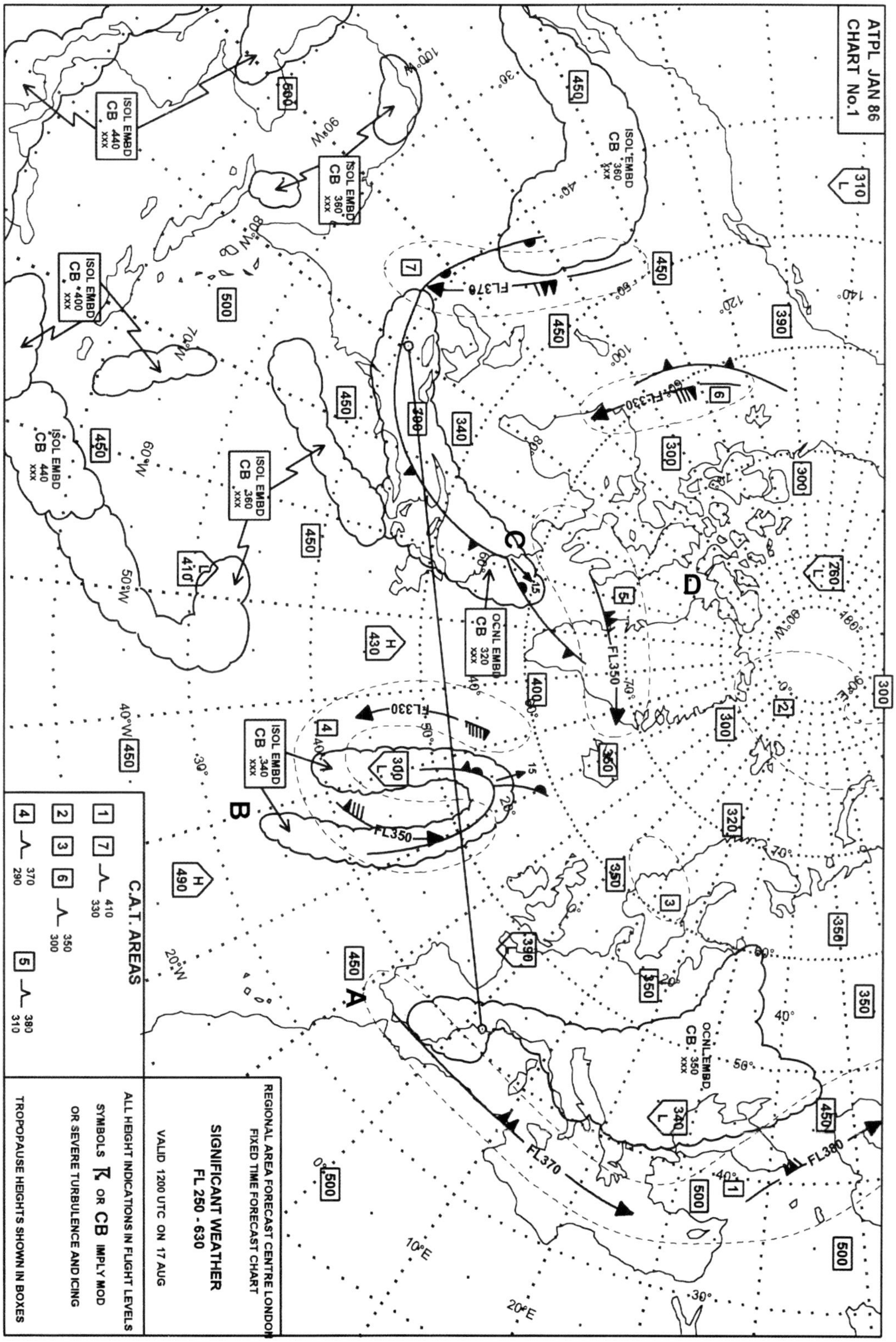
ATPL JAN 86
CHART No.1
REGIONAL AREA FORECAST CENTRE LONDON
FIXED TIME FORECAST CHART
SIGNIFICANT WEATHER
FL 250 - 630
VALID 1200 UTC ON 17 AUG
ALL HEIGHT INDICATIONS IN FLIGHT LEVELS
SYMBOLS OR CB IMPLY MOD
OR SEVERE TURBULENCE AND ICING
TROPOPAUSE HEIGHTS SHOWN IN BOXES
C.A.T. AREAS

Chart 1

UPPER WIND AND TEMPERATURE CHARTS

These charts are issued in conjunction with the significant weather chart and give spot winds from 700 hPa (FL100) up to 200 hPa (FL390). Spot values of wind and temperature are shown at regular intervals of latitude and longitude. The temperatures given are assumed to be negative unless prefixed by PS. The wind arrow symbology is exactly the same as that for the synoptic chart.

Chart 2 is also issued by London and is for Upper Wind and Temperature. Remember that the maximum wind is contained on the significant weather chart. The chart is for FL 340 and has the same validity as the upper wind chart. At the bottom is the time of issue — 1200 UTC on 16 August.

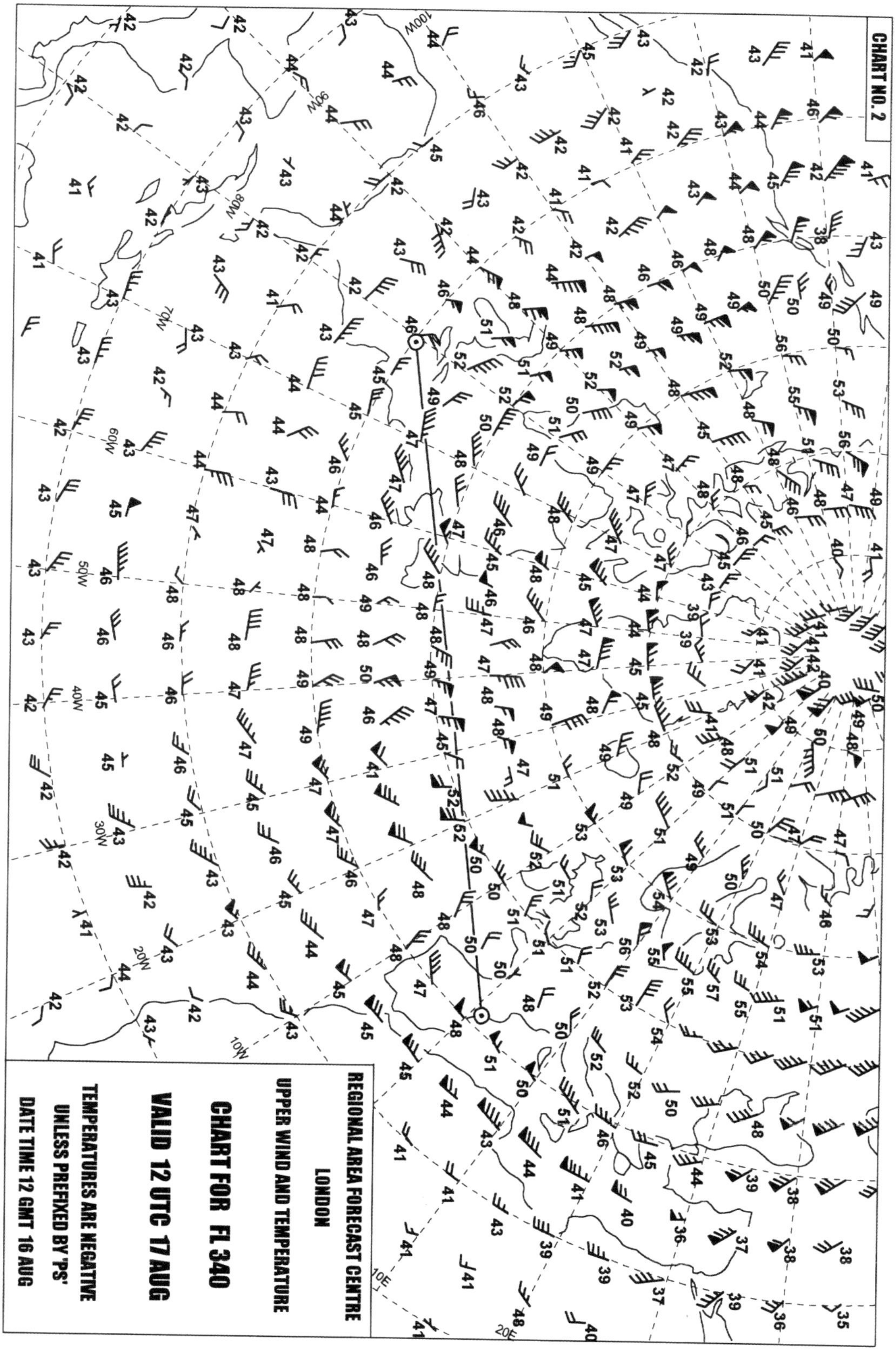
CHART NO. 2
REGIONAL AREA FORECAST CENTRE
LONDON
UPPER WIND AND TEMPERATURE
CHART FOR FL 340
VALID 12 UTC 17 AUG
TEMPERATURES ARE NEGATIVE
UNLESS PREFIXED BY 'PS'
DATE TIME 12 GMT 16 AUG

Chart 2

AVERAGING WIND VELOCITIES

When looking at wind charts, you need to apply some common sense. For instance, if there is an east/west track with a wind velocity of 310°/20 kt to the north and 270°/20 kt to the south, then the average wind might be 290°/20 kt. Numerical averaging is the common sense way of approaching the problem.

Example As well as taking spot winds and temperatures from specific points, winds and temperature need to be averaged over a route.

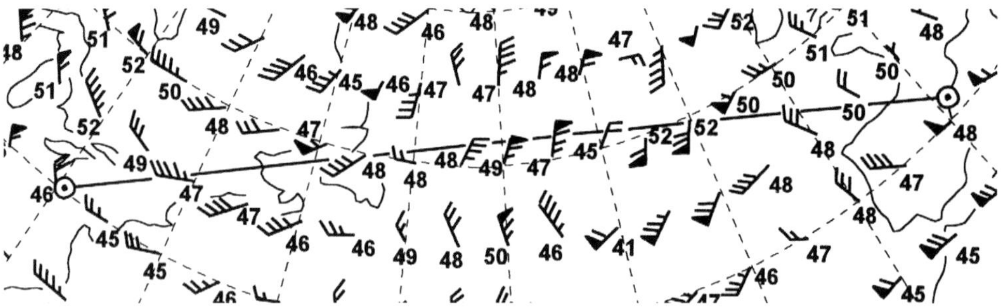

Because of the time limitations of the Flight Planning examination, the rule of KISS (keep it simple, stupid) applies.

Temperature

STEP 1 Along the route add up the temperatures and numerically average the sum total.

Temperature 48°C

The above system is quite a simple way of arriving at the mean temperature.

To average the wind velocity over a route is not as simple.

STEP 1 Look at the wind directions involved at approximately 10° spacing:

80W	320/20
70W	250/35
60W	240/50 (average between the two velocities spanning the track)
50W	270/15
40W	020/65
30W	020/20
20W	210/70
10W	290/30 (average between the two velocities spanning the track)
0E/W	270/50

The winds are predominantly westerly. Ignore the two north-easterly winds as they will distort the figures.

Direction 265°

STEP 2 For the speed, use the same principle as the direction. Give westerly winds a + configuration and easterly winds a – configuration.

Speed 25 knots

Time may mean that you are not able to make these calculations. If not, try to come to a sensible wind by inspection.

Chapter 15

Point of Equal Time, Point of Safe Return, and Radius of Action

INTRODUCTION

When flight planning, a pilot must be aware of the actions that need to be taken in an emergency. This will include the decision whether to:

- Return to the airport of departure, or
- Continue to the destination, or
- Fly to an alternate

This chapter shows how to calculate both the Point of Equal Time (Critical Point) and the Point of Safe Return (Point of No Return).

POINT OF EQUAL TIME

The Point of Equal Time (PET) is the point between two aerodromes from which it takes the same time to fly to either aerodrome.

For the still air case, the point of equal time is half way between the two aerodromes. This is not likely and so the PET is not half way between the two aerodromes. The calculation of the PET is based on a ratio of the groundspeed to the destination and groundspeed back to base. The TAS used for the calculation will depend upon whether the aircraft is to fly on:

- All engines, or
- One-engine inoperative

PET FORMULA

The PET is based on the statement that the time to destination is equal to the time to return to the aerodrome of departure.

You need to make certain assumptions for the calculation:

D is the total distance between airfields
X is the distance from the PET back to A
D-X is the distance to the destination (B)
H is the groundspeed home
O is the groundspeed to B

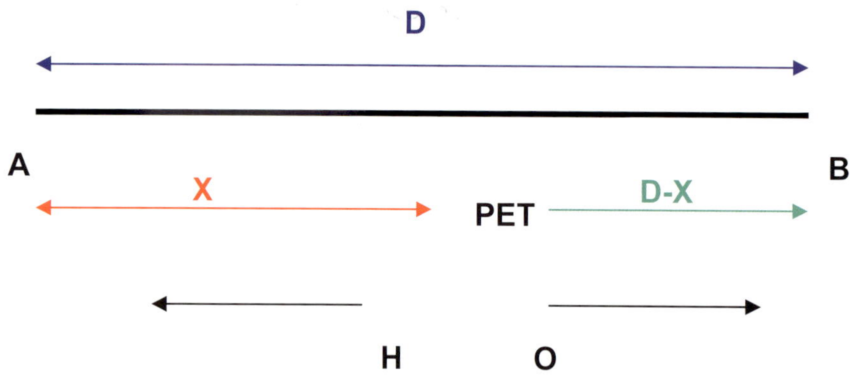

Time = Distance ÷ Groundspeed

PET is the point where time to destination is equal to the time to return to aerodrome of departure.

$$\text{Time to destination} = \frac{D-X}{O}$$

$$\text{Time to return} = \frac{X}{H}$$

$$\frac{X}{H} = \frac{D-X}{O}$$

$$\mathbf{X} = \frac{\mathbf{DH}}{\mathbf{O + H}}$$

X defines the distance of the PET from the departure.

Example Assume that points A and B are 600 nm apart.
TAS is 300 knots
Calculate the PET for the three conditions:

- Still air
- 50 knot headwind
- 50 knot tailwind

In the still air condition the PET must be halfway along the route **300 nm**

In the 50 knot headwind case
H = 350 knots
O = 250 knots

$$X = \frac{600 \times 350}{250 + 350} = \textbf{350 nm}$$

In the 50 knot tailwind case
H = 250 knots
O = 350 knots

$$X = \frac{600 \times 250}{350 + 250} = \textbf{250 nm}$$

To check that your calculation is correct you can check the time it takes to go to the B or return to A. In both cases it is one hour.

The wind effect moves the PET into wind.

PET Example 1	A – B	1240 nm
	TAS	340 KNOTS
	Wind Component	+20 knots outbound
PET Example 2	A – B	2700 nm
	TAS	450 KNOTS
	Wind Component	+50 knots outbound
PET Example 3	A – B	1400 nm
	TAS	270 KNOTS
	Wind Component	+40 knots outbound
PET Example 4	A – B	1120 nm
	TAS	210 KNOTS
	Wind Component	-35 knots outbound

ENGINE FAILURE PET

In most jet aircraft the loss of a power unit will cause "drift down". The aircraft descending to a pressure altitude that the power can sustain. Obviously there is now a decision to make as to whether the aircraft continues or returns.

Example **Using Example 2**

A – B	2700 nm
TAS	450 KNOTS
Wind Component	+50 knots outbound
PET from A	**1200 nm**
Time	**2 hours 24 minutes**

Consider the case of an engine failure, the TAS is most likely to be lower.

Assume a TAS of 360 knots and use the same details for Example 2:

H = 310 knots
O = 410 knots

$$X = \frac{2700 \times 310}{410 + 310} = 1162 \text{ nm}$$

PET from A **1162 nm**

With one engine inoperative, the wind has more effect, and the PET is removed further from mid-point than in the all engines operative case.

The aeroplane will fly with all engines operating until the engine failure. The reduced speed is used only to establish the one engine inoperative PET.

Therefore the time to the PET is the all engines groundspeed out.

A – B	1162 nm
GS	500 kt
Time	**2 Hours, 15 Minutes**

PET Example 5

A – B	2254 nm
Wind Component	-25 knots outbound
4 engine TAS	475 knots
3 Engine TAS	440 knots

Calculate the distance and time from A to the one engine out PET.

PET Example 6

A – B	1260 nm
Wind Velocity	020/35 knots
Course	040°T
4 engine TAS	480 knots
3 Engine TAS	435 knots

Calculate the distance and time from A to the one engine out PET.

PET Example 7

A – B	1700 nm
Wind Velocity	240/45 knots
Course	030°T
4 engine TAS	480 knots
3 Engine TAS	370 knots

Calculate the distance and time from A to the one engine out PET.

MULTI-LEG PET

Unfortunately most routes involve more than one leg. Therefore, you need to make multi-route calculations. Consider the route below.

TWO LEG PET

An aircraft is operating on the following route, what is the PET for one engine inoperative:

Route	Distance	Course	Wind Velocity
A – B	1025 nm	210	270/40
B – C	998 nm	330	280/20

4 Engine TAS 380 knots
3 Engine TAS 350 knots

STEP 1 Determine the groundspeed for:

B – C 334 knots
B – A 368 knots

STEP 2 Determine the times:

B – C 179 minutes
B – A 167 minutes

STEP 3 Because the time B – C is greater than the time B – A, the PET must be along B – C. To find the PET, the time of return must be equal to the time to travel to the destination.

Find the point along B – C (we will call this Point X) where the time to C is equal to the time B – A (167 minutes). This will leave us a distance to calculate the PET.

Groundspeed 334 knots
Point X 930 nm from C

STEP 4 The PET must lie between B and X. Distance BX is
998 – 930 = 68 nm

STEP 5 Using the PET formula calculate the PET for the 68 nm leg B – X
A return groundspeed is needed for X – B = 365 kts

$$\frac{68 \times 365}{334 + 365} = 35 \text{ nm from B}$$

A – PET is 1060 nm

STEP 6 To calculate the time to the PET calculate the four-engine time to B. The calculate the four engine time to the PET using the 35 nm calculated above.

A – B	**4 engine**	**172 min**
B – PET	**4 engine**	**6 min**
A – PET		**178 min**

THREE LEG PET

Consider the route below. Calculate the one-engine inoperative PET using the figures below.

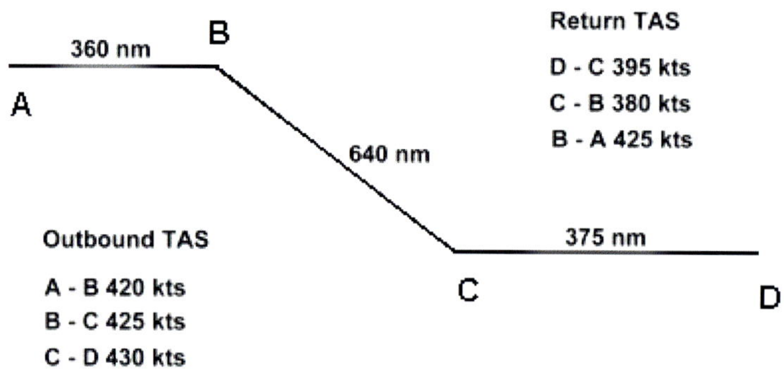

Outbound

Route	TAS	Wind Component	Groundspeed	Distance	Time
A – B	420	+30	450	360	48
B – C	425	+55	480	640	80
C – D	430	+20	450	375	50

Return

Route	TAS	Wind Component	Groundspeed	Distance	Time
D – C	395	-20	375	375	60
C – B	380	-60	320	640	120
B – A	425	-25	400	360	54

STEP 1 By inspection of the times it is obvious that the PET lies between B – C. Add all the outbound times together and halve them. 178 min total, therefore **89 minutes.** This would put you along leg B – C

STEP 2 To fly from B – A takes 54 minutes

To fly from C – D takes 50 minutes

If the times were equal, you could use the normal PET formula to calculate a PET between B – C. However, you have to equalise the times. Do this by determining how far the aircraft travels in four (54 – 50) minutes along the outbound leg.

Groundspeed 480 kts
Distance 32 nm

STEP 3 You now have the same time for the outbound as you do the inbound.

STEP 4 Now establish a PET for a revised distance of 608 nm (640 – 32)

$$\frac{608 \times 320}{320 + 480} = 243 \text{ nm}$$

Which makes the PET 243 nm from B

PET Example 8 Using the following data, calculate the distance and time to the one-engine inoperative PET for the following route:

4 Engine TAS 200 kts
3 Engine TAS 160 kts

Route	Course	Distance	Wind Velocity
A – B	115	170	180/20
B – C	178	110	230/30
C – D	129	147	250/15

PET Example 9 Using the following data, calculate the distance and time to the all engines operative PET for the following route:

TAS 175 kts

Route	TAS	Wind Component	Distance
A – B	175	-25 kt	450
B – C	175	-15 kt	430

PET Example 10 Using the following data, calculate the distance and time to the all-engines operative PET for the following route:

4 Engine TAS 250 kts

Route	Distance	Wind Component
A – B	252	-20
B – C	502	-5
C – D	310	+10

POINT OF SAFE RETURN

This is also known as the point of no return. The point of safe return (PSR) is the point furthest from the airfield of departure that an aircraft can fly and still return to base within its safe endurance.

Do not confuse the term "safe endurance" with the term "total endurance."

Total Endurance Is the time an aircraft can remain airborne, until the tanks are empty.

Safe Endurance Is the time an aircraft can fly without using the reserves of fuel that are required.

The distance to the PSR equals the distance from the PSR back to the aerodrome of departure.

Let:
E Safe endurance
T Time to the PSR
E – T Time to return to the aerodrome of departure
O Groundspeed to the PSR
H Groundspeed on return to the aerodrome of departure

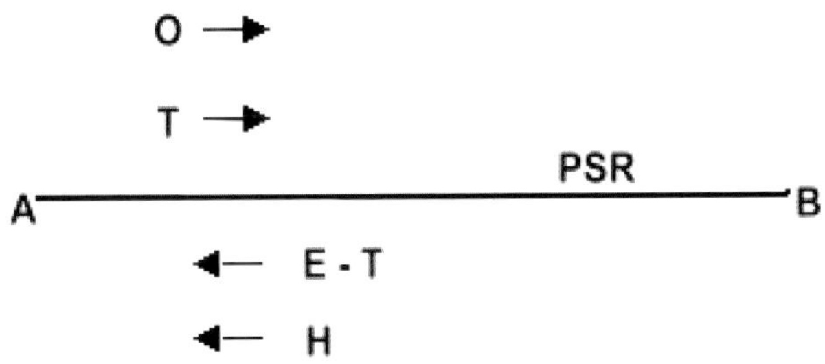

Time to the PSR T x O

Time to return to the aerodrome of departure (E – T) x H

(E – T) x H = T x O

$$T = \frac{EH}{O + H}$$

SINGLE LEG PSR

Given the following data, calculate the time and distance to the PSR.

TAS	220 kts
Wind Component	+45 kts
Safe Endurance	6 hours

$$T = \frac{360 \times 175}{175 + 265} = 143 \text{ minutes} = 632 \text{ nm}$$

PSR Example 1 Calculate the PSR given the following data:

A – B	800 nm
TAS	175 knots
Wind Component Outbound	-15 knots
Safe Endurance	5 hours

PSR Example 2 Calculate the PSR given the following data:

Fuel Available, excluding Reserve	21 240 lb
Fuel Consumption	3730 lb/hr
TAS Outbound	275 knots
TAS for Return Leg	285 knots
Wind Component Outbound	-35 knots

PSR Example 3 Calculate the PSR given the following data:

A – B	2200 nm
TAS	455 knots
Wind Component Outbound	-15 knots
Safe Endurance	6½ hours

MULTI-LEG PSR

Using the same principle as above, calculate the multi-leg PSR. Using the route below.

Route	Distance	Groundspeed		Time	
		Out	In	Out	In
A – B	300 nm	315 kts	440 kts	57 min	41 min
B – C	250 nm	375 kts	455 kts	40 min	33 min
C – D	350 nm	310 kts	375 kts	68 min	56 min

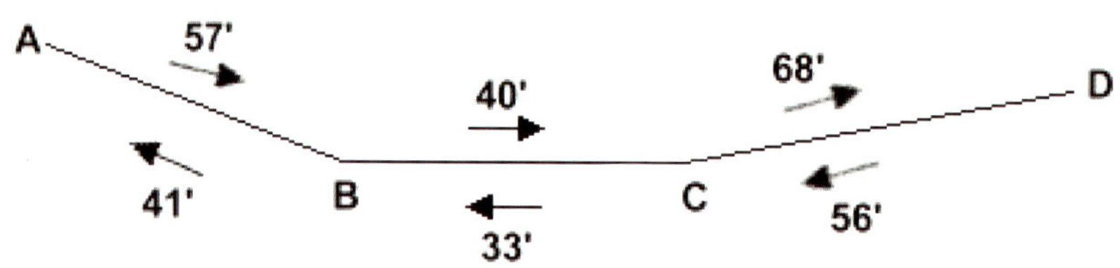

Safe Endurance is 210 minutes.

STEP 1 By inspection, determine on which leg the PSR is

Time A – B	57 min	Time B – C	40 min
Time B – A	<u>41 min</u>	Time C – B	<u>33 min</u>
	98 min		73 min
Total Time	171 min		

The Safe Endurance is 210 min
PSR must be on leg C to D

STEP 2 Remaining endurance is 39 min

Calculate the PNR for C – D using 39 min as the safe endurance.

$$T = \frac{39 \times 375}{310 + 375} = 21 \text{ min from C}$$

PSR Example 4 Calculate the time and distance to the PSR from A:

Route	Distance	TAS	Wind Component
A – B	520	200	-20
B – C	480	200	+6

Safe Endurance 6 hours 10 minutes

PSR Example 5 Calculate the time and distance to the PSR from A:

Route	Distance	TAS	Wind Component
A – B	410	250	-35
B – C	360	250	-25
C – D	200	250	-30

Safe Endurance 6 hours 10 minutes

PSR WITH VARIABLE FUEL FLOW

So far, the PSR has been given as a time. In the formula below the data is based upon the total fuel resolved into kg/nm.

Let:

D	Distance to the PSR	
F	Fuel available for the PSR	
FO	Fuel consumption out to the PSR	(kg/nm)
FH	Fuel consumption home from the PNR	(kg/nm)

The fuel used to get to the PSR plus the fuel used to get home from the PSR must equal the total fuel available (less reserves).

$$(d \times FO) + (d \times FH) = F$$

$$\mathbf{d = F \div (FO + FH)}$$

Example Given the following data, calculate the time to the PSR.

TAS	310 knots
Wind Component	+30 kt
Fuel Available	39 500 kg
Fuel Flow Out	6250 kg/hr
Fuel Flow Home	5300 kg/hr

STEP 1 Calculate the groundspeed out and the groundspeed home

Groundspeed Out	**340 kts**
Groundspeed Home	**280 kts**

STEP 2 Calculate the kg/nm for leg out and leg home

FO = 6250 ÷ 340 = 18.4 kg/nm

FH = 5300 ÷ 280 = 18.9 kg/nm

STEP 3 Calculate the time to the PSR

Distance	=	39 500 ÷ (18.4 + 18.9)
	=	1059 nm
Time	**=**	**187 minutes**

PSR Example 6 Given the following data, calculate the distance and time to the PSR

TAS Out	474 knots
Wind Component Out	-50 knots
Fuel Flow Out	11 500 lb/hr
TAS Home	466 knots
Wind Component Home	+70 knots
Fuel Flow Home	10 300 lb/hr
Flight Plan Fuel	82 000 lb
Reserves	12 000 lb

PSR Example 7 Given the following data, calculate the distance and time to the PSR

Leg Distance	1190 nm
TAS Out	210 knots
Wind Component Out	-30 knots
Fuel Flow Out	2400 kg/hr
TAS Home	210 knots
Wind Component Home	+30 knots
Fuel Flow Home	2000 kg/hr
Flight Plan Fuel	20 500 kg
Reserves	6000 kg

MULTI-LEG PSR WITH VARIABLE FUEL FLOW

In the previous multi-leg case, time out and time home were calculated on consecutive legs. In the variable fuel case, replace these figures by fuel out and fuel home and compare the total fuel burn.

Example Find the distance and time to the PSR from A given:

Route	Distance	TAS	Wind Component Out	Wind Component Home
A – B	270	480	-30	+35
B – C	340	480	-50	+55

Fuel Flow Out 11 900 kg/hr
Fuel Flow Home 11 650 kg/hr
Fuel Available 20 000 kg

STEP 1 Calculate the fuel A – B and B – A:

Time for Leg A – B **36.1 minutes**
Time for Leg B – A **31.5 minutes**

Fuel Used A – B **7160 kg**
Fuel Used B – A **6116 kg**
Fuel **13 276 kg**

STEP 2 Calculate the fuel remaining

20 000 – 13 276 = 6724 kg

STEP 3 The PNR is on B – C.

FO = 11 900 ÷ 430 = 27.7 kg/nm
FH = 11 650 ÷ 535 = 21.8 kg/nm

STEP 4 Calculate the distance for the PSR

D = 6724 ÷ (27.7 + 21.8)
D = 136 nm

The above distance is from B.

Total distance from A is **406 nm**

STEP 5 Calculate the time to the PSR

Time A – B **36.1 minutes**
Time B – PSR **18.2 minutes**
Time to PSR **54 minutes**

PSR Example 8 Given the following route, calculate the distance and time to the PSR assuming that the aircraft will return to A on 3 engines:

Route	Course	Distance	Wind Velocity
A – B	042	606	260/110
B – C	064	417	280/80
C – D	011	61	290/50

TAS 4 Engine	410 knots
TAS 3 Engine	350 knots
4 Engine Fuel Flow	3000 kg/hr
3 Engine Fuel Flow	2800 kg/hr
Fuel Available	12 900 kg

RADIUS OF ACTION

The radius of action is defined as:

"The distance to the furthest point from departure that an aircraft can fly, carry out a given flight, and return to its airfield of departure within the safe endurance"

The formula for radius of action is derived from the PSR formula and is:

$$E = \frac{E \times O \times H}{(O + H)}$$

Where: E is the safe endurance minus time on task

PET & PSR ANSWERS

PET Example 1	PET from A	584 nm
	Time	1 hour 37 minutes
PET Example 2	PET from A	1200 nm
	Time	2 hours 24 minutes
PET Example 3	PET from A	596 nm
	Time	1 hour 55 minutes
PET Example 4	PET from A	653 nm
	Time	3 hours 44 minutes
PET Example 5	PET from A	1191 nm
	Time	2 hours 39 minutes
PET Example 6	PET from A	679 nm
	Time	1 hour 32 minutes
PET Example 7	PET from A	760 nm
	Time	1 hour 28 minutes
PET Example 8	PET from A	221 nm
	Time	1 hour 11 minutes
PET Example 9	PET from A	488 nm
	Time	3 hours 14 minutes
PET Example 10	PET from A	540 nm
	Time	2 hour 16 minutes
PSR Example 1	PSR from A	163 minutes
	Distance	435 nm
PSR Example 2	PSR from A	195 minutes
	Distance	781 nm
PSR Example 3	PSR from A	201 minutes
	Distance	1477 nm
PSR Example 4	PSR from A	200 minutes
	Distance	611 nm
PSR Example 5	PSR from A	208 minutes
	Distance	760 nm

PSR Example 6	Distance	1510 nm
	Time	213 min
PSR Example 7	Distance	669 nm
	Time	223 min
PSR Example 8	Distance	765 nm
	Time	94 min

Chapter 16
Traffic Load

DEFINITIONS

Dry Operating Mass (DOM) is the total mass of an aeroplane ready for a specific type of operation excluding all usable fuel and traffic load. The mass includes:

- Crew and baggage
- Catering and removable passenger service equipment
- Potable water and WC chemicals
- Food and beverages

Traffic Load is the total mass of passengers, baggage, and cargo, including any "non-revenue" load.

Maximum Zero Fuel Mass (MZFM) is the maximum permissible mass of an aeroplane with no usable fuel.

Maximum Structural Take-Off Mass (MTOM) is the maximum permissible total aeroplane mass on landing under normal circumstances.

Maximum Structural Landing Mass (MLM) is the maximum permissible total aeroplane mass on landing under normal circumstances.

INTRODUCTION

The DOM varies as the role of the aircraft varies. The MZFM is determined by the airworthiness limits.

The MZFM is a stress limit and any extra weight above this limit is comprised of fuel only. Possibly the MZFM will limit the overall traffic load.

So Traffic Load = MZFM – DOM

The MTOM and MLM are other limitations on the traffic load under normal operating conditions.

MTOM comprises the DOM, route fuel at the start of the take-off run, and traffic load

MLM comprises of the DOM, the fuel remaining at touchdown, and the traffic load.

Consider all three limits separately to calculate the traffic load – the smallest of the three being the maximum traffic load. In any calculation, assume that the reserve fuel is unused unless otherwise stated.

Example Calculate the maximum traffic load given:

MTOM	210 000 kg
MLM	185 000 kg
MZFM	170 000 kg
DOM	125 000 kg
Fuel at Take-Off	45 000 kg
Fuel at Landing	10 000 kg

STEP 1 Calculate the stress limit Traffic Load
MZFM – DOM
170 000 – 125 000 = 45 000 kg

STEP 2 Calculate the TOM limit
MTOM – (DOM + Fuel at Take-Off)
210 000 – (125 000 + 45 000) = 40 000 kg

STEP 3 Calculate LM limit
MLM – (DOM + Fuel at Landing)
185 000 – (125 000 + 10 000) = 50 000 kg

STEP 4 The smallest figure is the maximum Traffic Load.
40 000 kg

Traffic Load Example 1 Given the following details, calculate the maximum traffic load:

MTOM	75 000 kg
MLM	60 000 kg
MZFM	55 000 kg
DOM	45 000 kg
Fuel at Take-Off	15 000 kg
Fuel Reserve	2000 kg

To calculate the maximum take-off weight when maximum payload is carried, use the MLM and the Route Fuel.

In the above case:

MLM 60 000 kg
Route fuel 13 000 kg (Fuel at take-off – reserve at landing)
Take-Off Weight when maximum payload is carried
73 000 kg

Traffic Load Example 2

Given the following details, calculate the maximum traffic load:

MTOM	54 000 kg
MLM	45 000 kg
MZFM	43 000 kg
DOM	31 000 kg
Fuel Consumption	1350 kg/hr
Fuel Reserve	2500 kg
TAS	220 kts
Wind Component	-20 kts
Leg distance	1200 nm

Traffic Load Example 3

Given the following details, calculate the maximum traffic load for this route:

MTOM	140 000 kg
MLM	95 000 kg
MZFM	85 000 kg
DOM	65 000 kg
Fuel Consumption	5000 kg/hr
Fuel Reserve	7000 kg
TAS	380 kts
Wind Component	+ 100 kts
Leg distance	4000 nm

Traffic Load Example 4

Given the following details, calculate the maximum traffic load for a leg of 1500 nm.

MTOM	115 000 kg
MLM	102 000 kg
MZFM	95 000 kg
DOM	60 000 kg
Fuel Consumption	6000 kg/hr
Fuel Reserve	5000 kg
Groundspeed	500 kts

Traffic Load Example 5

Given the following details, what is the extra fuel the aircraft can uplift:

MTOM	62 800 kg
MLM	54 900 kg
MZFM	51 300 kg
DOM	35 000 kg
Traffic Load	12 000 kg
Trip Fuel including Reserve	13 200 kg
Fuel Reserve	3100 kg

TRAFFIC LOAD ANSWERS

Traffic Load Question 1	10 000 kg
Traffic Load Question 2	11 500 kg
Traffic Load Question 3	20 000 kg
Traffic Load Question 4	32 000 kg
Traffic Load Question 5	2600 kg

Chapter 17

CAP 697 - Medium Range Jet Transport (MRJT) - Non-Normal Operations

GEAR DOWN FERRY FLIGHT *(CAP 697, PAGE 90)*

The graph is used in the same manner as Figure 4.3.1, giving the fuel required and time taken.

NON-NORMAL OPERATIONS QUESTION 1

Given the following, calculate the time and fuel required for a gear-down ferry flight:

Leg Distance	**600 nm**
Wind Component	**+ 50 knots**
Cruise Level	**FL 200**
Landing Weight	**40 000 kg**
ISA Deviation	**- 10°**

EXTENDED RANGE OPERATIONS *(CAP 697, PAGES 91 TO 95)*

The three figures in this section provide the planning information for:

- Critical fuel reserve
- Area of operation – diversion distance
- In flight diversion (LRC)

CRITICAL FUEL RESERVE – ONE ENGINE INOPERATIVE *(CAP 697, PAGE 92)*

Use the graph to determine the minimum fuel reserve at the PET (Critical Point). If this fuel reserve exceeds the planned fuel remaining at that point, then the fuel load must be adjusted accordingly.

Using Figure 4.7.1a, Page 92, note the factors upon which the figure is based:

- Emergency descent to 10 000 ft
- Level cruise at 10 000 ft
- 250 KIAS descent to 1500 ft
- One missed approach, approach, and land
- 5% allowance for wind errors
- Includes APU burn

The corrections at the bottom of the figure are similar to those encountered in earlier chapters.

- Increase fuel required by 0.5% for each 10° above ISA
- If icing conditions exist, increase fuel consumption by 20% to account for engine and wing anti-icing on and ice accumulation on unheated surfaces
- Allowance for performance deterioration is not included
- Compare the fuel required from this chart with the critical fuel reserves for two engines operative. Use the higher of the two.

NON-NORMAL OPERATIONS QUESTION 2

Calculate the critical fuel reserve needed if the following occurs:

Weight at Critical Point	50 000 kg
FL 100 Conditions	ISA + 20°C
Wind Component	- 50 knots
CP distance to Diversion	700 nm

CRITICAL FUEL RESERVE – ALL ENGINES OPERATIVE *(CAP 697, PAGE 93)*

Figure 4.7.1b is used in exactly the same manner as Figure 4.7.1a. The only change to this figure is that the extra allowance for anti-icing is reduced to 18%.

AREA OF OPERATION – DIVERSION DISTANCE (ONE-ENGINE INOPERATIVE) *(CAP 697, PAGE 94)*

Figure 4.7.2 relates to the region where an operator is authorised for ETOPS. The distance to a diversion airfield, from any point along route, must be flown within the approved time at the single-engine cruise speed. Assume still air and ISA conditions.

Example Given the following, find the diversion distance to a diversion airfield from any point on track:

Speed	0.74/290
Diversion Weight	45 000 kg
Time	120 minutes

STEP 1 Enter the chart from the left side with the speed and diversion weight.

STEP 2 Select the time along the top of the figure.

STEP 3 Where the two intersect, read off the diversion distance 792 nm.

NON-NORMAL OPERATIONS QUESTION 3

Fill in the following table with the diversion distance:

Speed	Diversion Weight	100 minutes	120 minutes	160 minutes
0.70/280	55 000 kg			
0.74/290	45 000 kg			
.74/330	60 000 kg			
LRC	35 000 kg			

IN-FLIGHT DIVERSION (LRC) – ONE ENGINE INOPERATIVE *(CAP 697, PAGE 95)*

Use figure 4.7.3 to calculate the fuel required and time from a point of diversion to an alternate aerodrome.

Use this figure in exactly the same way as Figures 4.3.1.

FUEL TANKERING AND FUEL PRICE DIFFERENTIAL *(CAP 697, PAGES 97/98)*

Fuel costs differ around the airports of the world. Sometimes, it is prudent to carry extra fuel when the fuel at the destination is much more expensive than at the airport of departure. This fuel can then be used for the return flight or next sector.

Two fuel tankering figures are given in CAP 697:

- Long Range Cruise
- 0.74 M Cruise

Use the tables to calculate the percentage of the surplus fuel used due to the increased weight of the aircraft.

Using the worked example for the LRC:

Trip Distance 1600 nm
Cruise Altitude FL 330
Landing Weight 42 500 kg

STEP 1 Enter the table with the Trip distance and move horizontally right to the pressure altitude line.

STEP 2 Move vertically down to the landing weight reference line. Then parallel the trade line to the landing weight.

Read off the surplus fuel burn **13.2%**.

Remember that this is a percentage of the extra fuel carried and not the total fuel carried.

Carry this figure forward to the Fuel Price Differential table on Page 98.

Example Use the figure of 13.2% calculated in the previous example. When the price at the departure aerodrome is 100 cents per gallon, what is the break-even price? (Ignore the worked example on the table).

STEP 1 Enter the table with the surplus fuel burn.

STEP 2 Move vertically to the 100 cents fuel price at departure airport. Next, move horizontally left to read the break even fuel price at the destination airport:

115 cents per gallon

NON-NORMAL OPERATIONS QUESTION 4

Calculate the % surplus fuel burn and the break even fuel price at the destination of an aircraft flying at LRC:

Cruise Altitude	FL 290
Trip Distance	2500 nm
Landing Weight	45 000 kg
Departure Airport Fuel Price	100 cents

NON-NORMAL OPERATIONS QUESTION 5

Calculate the % surplus fuel burn and the break even fuel price at the destination of an aircraft flying at LRC:

Cruise Altitude	FL 370
Trip Distance	1500 nm
Landing Weight	40 000 kg
Departure Airport Fuel Price	75 cents

NON-NORMAL OPERATIONS ANSWERS

Non-Normal Operations Question 1 Fuel 5500 kg
Time 1 hour 51 minutes

Non-Normal Operations Question 2 7070 kg

Non-Normal Operations Question 3

Speed	**Diversion Weight**	**100 minutes**	**120 minutes**	**160 minutes**
0.70/280	55 000 kg	630	752	997
0.74/290	45 000 kg	663	792	1050
.74/330	60 000 kg	656	783	1035
LRC	35 000 kg	608	728	965

Non-Normal Operations Question 4 Surplus Fuel Burn 19.8%
Break Even Fuel Price 125 cents

Non-Normal Operations Question 5 Surplus Fuel Burn 14.7%
Break Even Fuel Price 88 cents